Starch: Progress in Food Applications

Germán Ayala Valencia
Editor

Starch: Progress in Food Applications

Springer

Editor
Germán Ayala Valencia
Department of Chemical and Food Engineering
Federal University of Santa Catarina
Florianópolis, Brazil

ISBN 978-3-031-98731-1 ISBN 978-3-031-98732-8 (eBook)
https://doi.org/10.1007/978-3-031-98732-8

This Springer imprint is published by the registered company Springer Nature Switzerland AG
The registered company address is: Gewerbestrasse 11, 6330 Cham, Switzerland

This book is dedicated especially to my daughter Liz and my wife Talita.

Germán Ayala Valencia, Ph.D.
Editor

Preface

Starch is a natural ingredient broadly used in the food industry due to its thickening, binding, gelling, baking, and texturing properties. In recent years, starches have been used to develop new structures with potential food applications, such as aerogels or ingredients stabilizing bioactive compounds and proteins. Similarly, the knowledge about the starch structure, its modification, and properties has allowed the production of ingredients with a low glycemic index, 3D food printing, and plant-based foods. This book aims to bring together recent advances in the food application of starches. Thus, Chap. 1 analyzes the starch architecture. The advances in the production of starch aerogels and starch ingredients stabilizing bioactive compounds and proteins have been reviewed in Chaps. 2, 3 and 4. The structuration of starches and its impact on digestibility and production of ingredients with low glycemic index are reviewed in Chaps. 5 and 6. Chapters 7, 8, and 9 analyze the use of starches in 3D food printing, the production of modified starches by enzymatic methods, and the use of starch to manufacture plant-based foods, respectively. Finally, Chap. 10 reviews the use of starch in clean label foods.

I would like to sincerely thank the contributors for their insights into *Starch: Progress in Food Applications*.

Florianópolis, SC, Brazil — Germán Ayala Valencia

Contents

Editor and Contributors

About the Editor

Germán Ayala Valencia received his Ph.D. in food engineering in 2017 from the University of São Paulo, Brazil. He is a professor at the Department of Chemical and Food Engineering in the Federal University of Santa Catarina, Florianópolis, Brazil. He was the guest editor of two special issues published by Starch—Stärke (11–12/2021 and 1–2/2024). Currently, he is a Board Member of the *Journal of Packaging Technology and Science* and editor of the books *Natural Additives in Foods* and *Starch Nanomaterials and Food Applications*, both published by Springer Nature. He mainly works in the food science and technology area with an emphasis on natural polymers, food packaging, natural pigments, nanotechnology, agro-industrial waste, and stabilization of bioactive compounds.

Contributors

Edith Agama-Acevedo Instituto Politécnico Nacional, CEPROBI, Yautepec, Mexico

J. Alvarez-Ramirez Departamento de Ingeniería de Procesos e Hidráulica, Universidad Autónoma Metropolitana-Iztapalapa, Iztapalapa, Mexico

Renan Nascimento Araújo São Carlos Institute of Chemistry (IQSC), University of São Paulo – USP, São Carlos, SP, Brazil

Liliana Avila-Martín Departamento de Ingeniería Química y Ambiental, Universidad Nacional de Colombia, Bogotá D.C., Colombia

Sara Kierulff Balabram São Carlos Institute of Chemistry (IQSC), University of São Paulo – USP, São Carlos, SP, Brazil

Luis A. Bello-Perez Instituto Politécnico Nacional, CEPROBI, Yautepec, Mexico

Nicolas M. Bogdanoff Universidad Paraguayo Alemana. Facultad de Ciencias de la Ingeniería, San Lorenzo, Paraguay

Wilson Daniel Caicedo Chacon Department of Chemical and Food Engineering, Federal University of Santa Catarina, Florianópolis, SC, Brazil

Ingeniería de Alimentos, Fundación Universitaria Agraria de Colombia, Bogotá, Colombia

Raul Remor Dalsasso Department of Chemical and Food Engineering, Federal University of Santa Catarina, Florianópolis, SC, Brazil

Raquel da Silva Simão Department of Chemical and Food Engineering, Federal University of Santa Catarina, Florianópolis, SC, Brazil

Samile Bezerra de Aguiar São Carlos Institute of Chemistry (IQSC), University of São Paulo – USP, São Carlos, SP, Brazil

Fátima Díaz Universidad Nacional de Asunción, Facultad de Ciencias Químicas, Departamento de Ingeniería y Tecnología de Alimentos, San Lorenzo, Paraguay

Omayra B. Ferreiro Universidad Nacional de Asunción, Facultad de Ciencias Químicas, Departamento de Ingeniería y Tecnología de Alimentos, San Lorenzo, Paraguay

Universidad Nacional de Asunción, Facultad Politécnica, Grupo de Investigación en Bio y Materiales, San Lorenzo, Paraguay

Valeria Cedeno Fierro Facultad de Ingeniería de Administración, Universidad Nacional de Colombia, Palmira, Colombia

J. Figueroa-Flórez Facultad de Ingeniería, Programa de Ingeniería Agroindustrial, Universidad de Sucre, Sincelejo, Colombia

Facultad de Ingeniería y Administración, Programa de Ingeniería Agroindustrial, Universidad Nacional de Colombia, Palmira, Colombia

Maria A. Hernandez-Aguirre Instituto Politécnico Nacional, CEPROBI, Yautepec, Mexico

L. Hernández-Vanegas Facultad de Ingeniería, Programa de Ingeniería Agroindustrial, Universidad de Sucre, Sincelejo, Colombia

Pedro Augusto Invernizzi Sponchiado São Carlos Institute of Chemistry (IQSC), University of São Paulo – USP, São Carlos, SP, Brazil

Stanislau Bogusz Junior São Carlos Institute of Chemistry (IQSC), University of São Paulo – USP, São Carlos, SP, Brazil

Gabriel Coelho Leandro Department of Chemical and Food Engineering, Federal University of Santa Catarina, Florianópolis, SC, Brazil

Eduardo Leonarski Department of Chemical and Food Engineering, Federal University of Santa Catarina, Florianópolis, SC, Brazil

Bianca Chieregato Maniglia São Carlos Institute of Chemistry (IQSC), University of São Paulo – USP, São Carlos, SP, Brazil

Juan Felipe Grisales Mejía Chemistry Department, Laboratory of Advanced Analytical Techniques in Natural Products (LATNAP), Universidad de los Andes, Bogotá, Colombia

Perla Osorio-Diaz Instituto Politécnico Nacional, CEPROBI, Yautepec, Mexico

Jairo E. Perilla Departamento de Ingeniería Química y Ambiental, Universidad Nacional de Colombia, Bogotá D.C., Colombia

Mikaela Rajchman Institute of Food Science Research - CIAL (CSIC-UAM), Madrid, Spain

J. Salcedo-Mendoza Facultad de Ingeniería, Programa de Ingeniería Agroindustrial, Universidad de Sucre, Sincelejo, Colombia

T. Serna-Fadul Facultad de Ingeniería, Programa de Ingeniería Agroindustrial, Universidad de Sucre, Sincelejo, Colombia

Guilherme Dallarmi Sorita Department of Chemical and Food Engineering, Federal University of Santa Catarina, Florianópolis, SC, Brazil

Larissa Tessaro São Carlos Institute of Chemistry (IQSC), University of São Paulo – USP, São Carlos, SP, Brazil

E. J. Vernon-Carter Departamento de Ingeniería de Procesos e Hidráulica, Universidad Autónoma Metropolitana-Iztapalapa, Iztapalapa, Mexico

Abbreviations

AC	Amylose content
AM	Amylose
BA-PC	Blueberry anthocyanin-phycocyanin
CAD	Computer-aided design
CD	Crystallinity degree
CGTase	Cyclodextrin glucanotransferase
CHMT	Continuous heat-moisture treatment
CLS	Clean label starch
CLSM	Confocal laser scanning microscopy
CM	Chemically modified
DHT	Dry heat treatment
DIW	Direct ink writing
DMSO	Dimethyl sulfoxide
DSC	Differential scanning calorimetry
FDA	Federal Drug Administration
FTIR	Fourier-transform infrared
GA	Gallic acid
GCS	Granular cold-swelling
GI	Glycemic index
GIP	Glucose-dependent insulinotropic polypeptide
HAMS	High-amylose maize starch
HAWF	High-amylose wheat flour
HM	High moisture
HMT	Heat-moisture treatment
HMW-GSs	High-molecular-weight glutenin subunits
HPAEC	High-performance anion-exchange chromatography
HPMC	Hydroxypropyl methylcellulose
HPP	High hydrostatic pressure
IDDSI	Dysphagia Diet Standardization Initiative
LOP	Logarithm of the slope
Ni	Not informed

NMR	Nuclear magnetic resonance
OSA	Octenyl succinic anhydride
pI	Isoelectric point
PPI	pea protein isolate
PS	Pore size
RD	Retrogradation degree
RDS	Rapidly digestible starch
RHMT	Repeated heat-moisture treatment
RPH	Rice protein hydrolysates
RS	esistant starch
RVA	Rapid visco analyzer
RVA	Rapid visco-analysis
SAXS	Small-angle X-ray scattering
SC	Supercritical
SCFAs	Short-chain fatty acids
SCo	Starch concentration
SDS	Slowly digestible starch
SEM	Scanning electron microscopy
SFF	Solid freeform fabrication
SSA	Specific surface area
Tc	Conclusion temperature
TEM	Transmission electron microscopy
T_g	Glass transition temperature
T_{gel}	Gelatinization temperature
TI	Thermally inhibited
To	Onset temperature
Tp	Peak temperature
T_{ret}	Retrogradation temperature
WHO	World Health Organization
XRD	X-ray diffraction
ΔH	Gelatinization enthalpy
ρ	Density
ϕ	porosity

Chapter 1
Starch Architecture

Jairo E. Perilla and Liliana Avila-Martín

1.1 Introduction

Starch is a natural polysaccharide synthesized by plants as an energy reserve during dormancy, growth, or stress, serving as the largest carbohydrate source in nature [1]. It is stored in granules within plant tissues such as leaves, roots, seeds, flowers, and stems, with its accumulation regulated by environmental signals, including light [2]. The molecular structure of starch varies among botanical sources due to differences in amylose and amylopectin composition, which influence its physicochemical properties and determine its suitability for industrial and nutritional applications [3, 4]. In crops such as potatoes, starch biosynthesis is controlled by enzymatic pathways, including starch synthases and branching enzymes, which regulate granule formation and the amylose-to-amylopectin ratio, directly affecting functionality [3]. This physicochemical versatility makes starch an essential ingredient in food products as a thickener, stabilizer, and texturizer, while also playing a crucial role in pharmaceuticals, textiles, and biodegradable packaging [5, 6].

With increasing sustainability concerns and industrial demand, research on non-conventional starch sources has expanded to include alternatives such as plantain (*Musa × paradisiaca*) and achira (*Canna indica*) [5, 7, 8]. These alternative sources possess distinctive structural and functional properties, offering promising solutions for food and non-food applications. Modification techniques, such as the incorporation of fatty acids like stearic acid, enhance starch's thermal stability and mechanical properties, making it a viable component in biodegradable materials [4]. Additionally, advancements in extraction technologies, including enzymatic and ultrasound-assisted methods, have been explored to optimize starch yield and

J. E. Perilla (✉) · L. Avila-Martín
Departamento de Ingeniería Química y Ambiental, Universidad Nacional de Colombia, Bogotá D.C., Colombia
e-mail: jeperillap@unal.edu.co

G. Ayala Valencia (ed.), *Starch: Progress in Food Applications*,
https://doi.org/10.1007/978-3-031-98732-8_1

minimize environmental impact [6]. Recent developments in starch applications extend to sustainable construction materials, exemplified by "StarchCrete," a lightweight composite with potential for environmentally friendly building solutions [9]. The ongoing study and development of starch-based materials contribute to a more sustainable industrial landscape by offering innovative alternatives across various sectors.

A deeper understanding of the correlation between the physicochemical and functional properties of native starches provides valuable insight into the performance of starch-based products from different botanical sources. Physically, starch exists in discrete granules and consists of two primary molecules: amylose and amylopectin. The quality and composition of starches are intrinsically linked to the plant source from which they are obtained, influencing their applications in various industries and technological advancements.

1.2 Amylose and Amylopectin

Starch is a polysaccharide primarily composed of two types of macromolecules: amylose and amylopectin as seen in Fig. 1.1. Amylose is a linear polymer of glucose units linked by α-(1→4) bonds. Although mostly linear, it can exhibit a small percentage of branching with α-(1→6) linkages, which make up less than 2% of the total links [10–12]. Its molecular weight varies depending on the botanical source, ranging from 10,000 to 1,000,000 Da, with degrees of polymerization fluctuating between 100 and 10,000 glucose units [13, 14]. The linear configuration of amylose

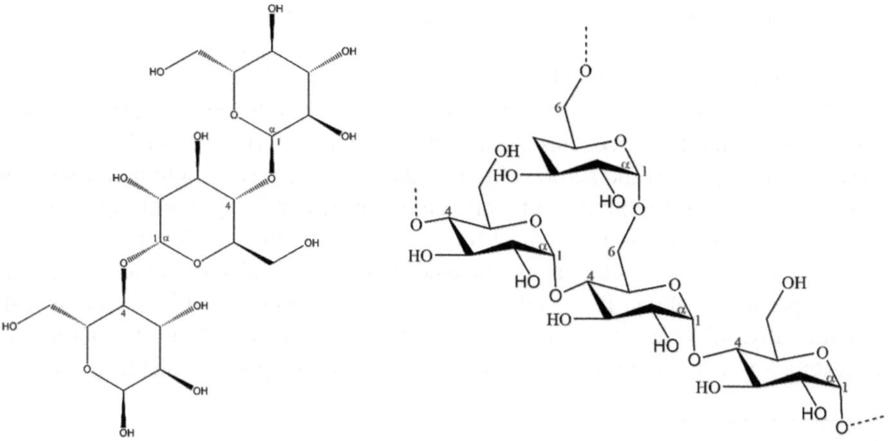

Fig. 1.1 Structural representation of starch components: amylose (left) and amylopectin (right). Amylose is a primarily linear polysaccharide composed of α-(1→4) linked glucose units, contributing to gel formation and film strength. Amylopectin, in contrast, is a highly branched polysaccharide with α-(1→4) linked glucose chains and α-(1→6) branch points, responsible for starch viscosity and swelling properties

facilitates the formation of simple or double helices stabilized by hydrogen bonds, which influences key starch properties such as density and crystalline organization. These helices can host small molecules, such as lipids and iodine, facilitating the formation of type V complexes crucial for the thermal properties of starch and its resistance to enzymatic digestion [11, 12]. Moreover, amylose contributes to the formation of B-type crystalline structures, characteristic of starches high in this molecule [13].

In contrast, amylopectin is a highly branched polysaccharide. α-(1→4) bonds form the backbone, while branching points consist of α-(1→6) linkages occurring approximately every 20–30 glucose units. Its molecular weight is considerably larger than that of amylose, reaching up to 100 million Da, making it one of the largest biopolymers in nature [11, 12, 15]. The branched structure of amylopectin generates three types of chains: A (unbranched and short, 12–24 glucosyl units), B (branched and longer, 45–55 glucosyl units), and C (linking the clusters of the molecule). This organization allows the formation of double helices that group into semicrystalline regions, responsible for the A, B, and C crystalline patterns observed in starch granules.

The amylose to amylopectin ratio varies depending on the botanical source, with amylopectin constituting between 65% and 85% of the starch and amylose between 15% and 35% [16]. The structure and properties of these two polymers directly influence starch characteristics such as gelatinization, thermal stability, and water absorption capacity.

The measurement of starch molecular weight is complex due to its mixed nature, combining amylose and amylopectin. Various techniques, such as chromatography and light scattering, are used to determine the starch molecular weight. For instance, the molecular weight of amylose can range between 1×10^5 and 1×10^6 g/mol, while amylopectin's molecular weight varies widely, reaching up to 100 million Da in some cases. For potato starch, the molecular weight is found between 23 and 37×10^6 g/mol [17], whereas for plantain, it is between 2.88 and 3.14×10^8 g/mol [18]. Additionally, cassava starch has a molecular weight of 25.52×10^5 g/mol [19]. Differences observed in studies can be attributed to variations in sample preparation methods, molecular standards, and chromatographic conditions, making direct comparison of results from different studies difficult [20].

1.3 Granular Structure and Crystallinity

Starch forms a macrostructure denominated granules. The granules' size and shape depend on starch's botanical origin. For example, Maize starch typically has small, polygonal granules ranging from 2 to 20 μm. Potato starch granules are much larger (15–100 μm) and have an oval to irregular shape with distinctive growth rings. Cassava starch granules are round or oval, with sizes ranging from 5 to 35 μm, contributing to its smooth texture and high paste clarity. Plantain starch has an elongated to oval shape with a granule size distribution between 10 and 50 μm, while achira starch exhibits large, irregularly shaped granules (20–120 μm). Depending on the botanical origin and the moisture availability during granule formation, the two glucans that constitute starch—amylose and amylopectin—are organized into amorphous and crystalline regions, along with other starch components [11, 12].

Due to its linear structure with α-(1→4) linkages, Amylose tends to form simple or double helices stabilized by hydrogen bonds. These helices can interact with lipids, such as fatty acids and phospholipids, forming helical complexes known as type Vh crystals. These complexes exhibit thermal stability and resistance to digestion, making them relevant in industrial and food applications [11, 12]. On the other hand, amylopectin, with its branched structure, facilitates the formation of denser double helices that group into semicrystalline regions. These helices are essential for the organization of A-type (monoclinic) and B-type (hexagonal) crystalline structures, depending on chain length and granule hydration (Fig. 1.2) [14, 21].

The amylose to amylopectin ratio directly influences the density and stability of starch crystalline structures. High amylose levels limit crystallinity due to its lower capacity to form well-ordered double helices, while a high amylopectin content favors the formation of A or B-type crystals, depending on environmental conditions and granule type. A-type starches, characteristic of cereals like corn, display a compact packing with less inter-helical space, granting greater crystalline density and water resistance. Cassava and other roots also form A-type crystals. Contrary to

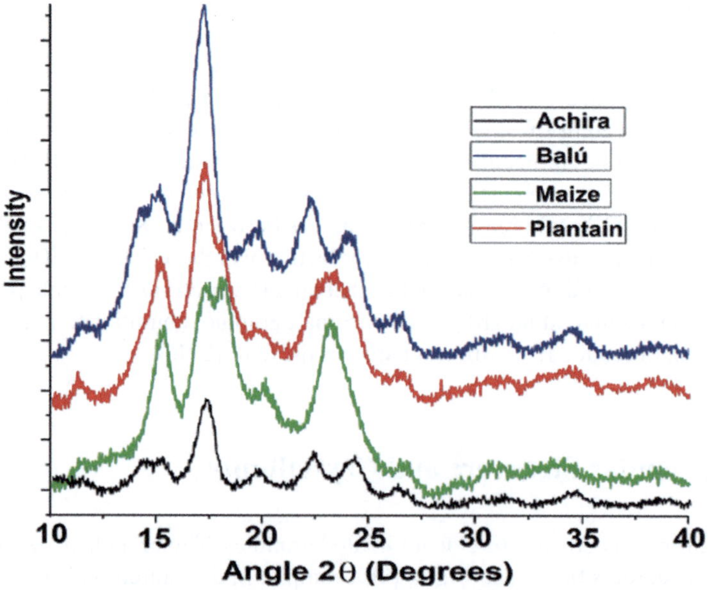

Fig. 1.2 Types of crystalline structures in starch. X-ray diffraction (XRD) patterns of starch granules from achira, balú, maize, and plantain. Maize exhibits A-type crystallinity with strong, well-defined peaks, while achira and balú show B-type crystallinity with broader, more hydrated diffraction patterns. Plantain starch presents CA-type crystallinity, an intermediate structure between A- and B-type arrangements

A-type crystals, B-type starches, common in tubers like potato and achira, have more spaced helices, allowing higher water content in amorphous zones. C-type starches, typical of legumes like plantain, combine characteristics of A and B types [10, 13]. In terms of diffraction, A-type starches exhibit higher relative crystallinity with peaks at 14.8°, 17.2°, and 23.1° (2θ), while C-type starches present a combination of A and B polymorphs. B-type starches, such as those from achira, avocado, and balú, show lower relative crystallinity with peaks at 15.5°, 17.28°, 19.7°, 22.3°, and 24.1° (2θ), as indicated by various studies [22–25].

Lipids associated with starch, particularly phospholipids, play a fundamental role in stabilizing amylose simple helices and forming helical complexes. These amylose-lipid complexes not only affect crystalline organization but also enhance thermal stability and limit starch retrogradation. Granular proteins, such as granule-bound starch synthase, regulate the amylose to amylopectin ratio, influencing the biosynthesis and arrangement of crystalline and amorphous regions. These interactions between lipids, proteins, and structural carbohydrates determine starch functional properties such as gelatinization, retrogradation, and resistance to enzymatic digestion [16, 26].

The hierarchical organization of starch is also reflected in blocklet formation, intermediate structures composed of crystalline and amorphous lamellae. Blocklets, varying in size depending on the botanical source, are primarily composed of packed double helices of amylopectin in crystalline regions. Amorphous areas contain longer and more disordered chains, influenced by the presence of lipids and proteins. A higher amylopectin proportion promotes structural regularity in blocklets, while high amylose levels introduce discontinuities due to its lower propensity to form well-ordered helices [12, 21].

In summary, the interaction between amylose, amylopectin, lipids, and proteins determines the formation and stability of starch crystalline structures. These interactions not only affect the physical and functional properties of starch but are also critical for its application in various industries, from food to bio-based materials [11, 27].

Genetics plays a crucial role in the starch proportion and granular structure. Mutations in genes coding for key enzymes, such as branching enzymes and debranching enzymes, significantly modify the composition and molecular architecture of starch chains. For instance, waxy starches, which contain almost exclusively amylopectin, exhibit highly organized structures and high crystallinity due to mutations in granule-bound starch synthase activity, whereas starches with high amylose content have more amorphous structures, lower crystallinity, and greater resistance to enzymatic hydrolysis [10, 27]. These structural variations also influence starch functional properties such as retrogradation and the applicability of starch in controlled release coatings [12]. In this regard, genetic regulation of starch is key to adapting it to specific needs in sectors such as food, pharmaceutical, and biomaterials (Table 1.1).

Table 1.1 Comparative characteristics of starch granules from different botanical sources

Botanical origin	Granule size (μm)	Morphology	Crystal type	Gelatinization temperature (°C)	References
Potato	22.1–45.0	Irregular and lenticular	B-type	63.8–68.1	[28]
High-amylose maize	<10	Polygonal. Varying based on amylose content	N. A	>75	[29]
Wheat	<20	Polygonal, disk-shaped	A-type	61	[30]
Cassava	4–20	Spherical or hemispherical	C-type, progresses to A- or B-type	70–74	[31]
Sweet potato	0.1–100	Round, polygonal, oval, and semi-oval shapes	CA to B to CC-type, harvesting dependent	66.0–75.6	[32]

1.4 Starch Gelatinization

The gelatinization and retrogradation of starch are critical processes that significantly impact its functionality. Gelatinization involves the thermal transition of starch granules from a birefringent semi-crystalline structure to an amorphous state. Gelatinization is essential for improving starch digestibility and its interaction with other components. Conversely, retrogradation refers to reorganizing gelatinized starch molecules into crystalline structures during cooling and storage. Retrogradation has a significant impact on the texture and stability of starch-based products. Understanding the factors influencing these processes becomes essential for optimizing the applications of starches in food science and other industrial applications [33].

1.4.1 Gelatinization: A Transition from Order to Disorder

Gelatinization is a thermal process in which starch granules transition from a semi-crystalline to an amorphous state when heated in the presence of water. This transformation is driven by a series of molecular and structural changes within the granules and occurs in the following steps:

- Initial water absorption: Starch granules comprise crystalline and amorphous regions. Upon heating in the presence of water, water molecules initially diffuse in the amorphous regions of the granules. This hydration process begins to soften the structure, enabling water to interact with the polymers and initiate granule swelling. Hydration occurs preferentially in the amorphous areas, where molecular packing is less dense.

- Swelling of Granules: As the temperature increases, water molecules disrupt the weak hydrogen bonds in the amorphous regions of the starch granules. This disruption leads to significant swelling as granules take in more water. Amylose, being less entangled and more linear, starts to mobilize and leach out of the granules. Amylopectin, with its branched and highly crystalline structure, resists swelling initially, requiring additional thermal energy to undergo disruption.
- Disruption of Amorphous Regions: Continued heating further destabilizes the amorphous regions. The hydrogen bonds that maintain structural integrity are progressively weakened, resulting in increased molecular mobility. These effects initiate the destabilization of the crystalline regions, which are predominantly composed of amylopectin double helices.
- Loss of Crystalline Order: In a temperature range, specific for each starch, the crystalline regions undergo a phase transition losing the birefringent pattern and conducting the structure to an amorphous state. The hydrogen bonds holding the amylopectin double helices together are disrupted, and crystallites may no longer exist under these conditions. Besides the loose of birefringence, this transition is also associated with an endothermic event that resembles the melting of crystals.
- Leaching of amylose: As the crystalline order is lost, amylose molecules, which are more mobile due to their linear structure, are released from the granules into the surrounding water. This leaching of amylose is critical for the development of viscosity in starch-based systems, as it facilitates the formation of a viscous matrix.
- Granule Rupture and Formation of a Viscous Paste: In the final stage of gelatinization, the structural integrity of the starch granules is lost entirely. The disrupted amylopectin chains, along with the leached amylose, interact with water to form a cohesive and viscous paste. This matrix contributes to the unique functional properties of gelatinized starch in food and industrial applications.

1.4.2 Characterization of the Gelatinization Process

Since gelatinization is a temperature-driven hydration process, differential scanning calorimetry (DSC) stands out as one of the most effective tools for evaluating and characterizing this phenomenon. Under a controlled heating rate, for example, 10 K/min, DSC provides essential thermal parameters, including the onset temperature (To), peak temperature (Tp), and gelatinization enthalpy (ΔH) as shown in Fig. 1.3. These parameters quantify the energy required to disrupt the crystalline structures within starch granules, offering a detailed understanding of the gelatinization process and its thermal transitions [34]. Usually, the peak temperature is reported as the gelatinization temperature, and higher ΔH values indicate a greater degree of crystallinity, often associated with amylopectin-rich starches like potato or high-amylopectin maize [35].

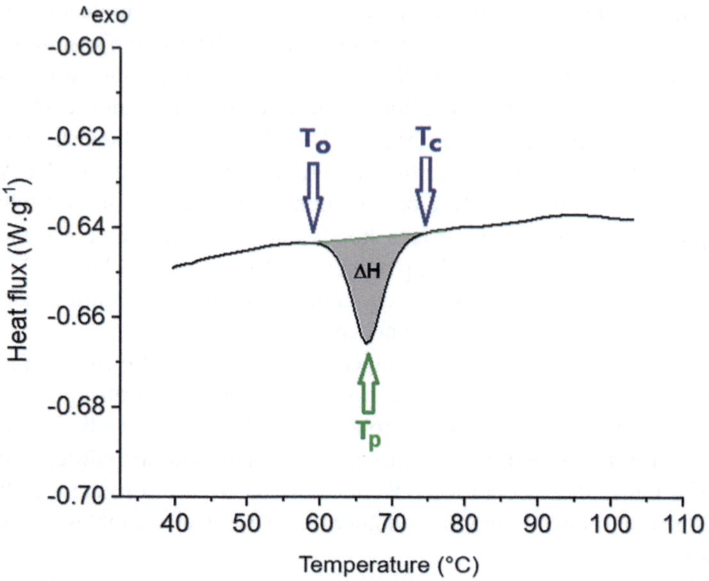

Fig. 1.3 Differential Scanning Calorimetry (DSC) thermogram illustrating the starch gelatinization. The endothermic peak represents heat absorption during the process. Key thermal parameters include T_0 (onset temperature, blue) marking the transition's beginning, T_p (peak temperature, green) indicating maximum heat absorption, and T_c (conclusion temperature, blue) where the transition ends. The shaded area (ΔH) corresponds to the enthalpy change associated with the event

During gelatinization, the system goes from a dispersion of starch granules in water with a viscosity characterized by the viscosity of the water and the concentration of the granules, to a gel state with a rheological behavior resembling a polymer solution, with non-Newtonian behavior and elastic effects. Therefore, rheological techniques, such as rapid visco-analysis (RVA) and visco-amylography, are used to determine the gelatinization temperature by tracking changes in viscosity during gelatinization. These methods measure pasting temperature, peak viscosity, hold viscosity, and breakdown viscosity, reflecting the mechanical stability of granules during heating and stirring [36].

In a typical RVA (Rapid Visco Analyzer) graph as shown in Fig. 1.4, the viscosity of a starch suspension is plotted against time, capturing its behavior under controlled heating, holding, and cooling cycles. The process begins with a flat baseline at the initial phase (I), where the starch granules are dispersed in water but remain unaltered. This phase represents the suspension's low viscosity as minimal hydration occurs, establishing a steady starting point for the analysis.

As the temperature increases, the line on the graph begins to slope upward, indicating the onset of starch gelatinization (II). Water penetrates the amorphous regions of the granules, causing them to swell and absorb water. This swelling increases the viscosity gradually until the crystalline regions of amylopectin melt, marking the gelatinization peak. At this point, the line reaches its maximum, which corresponds

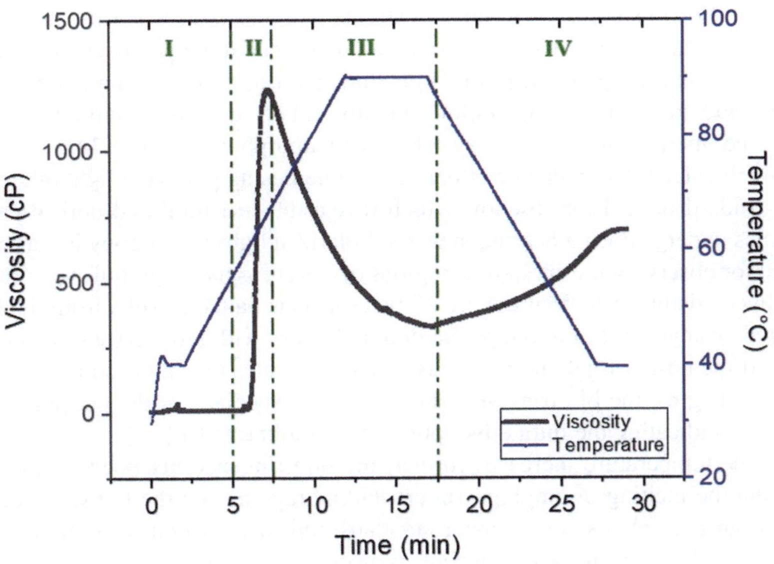

Fig. 1.4 Rapid Visco Analyzer (RVA) viscosity profile of native potato starch, illustrating its pasting behavior. The graph presents viscosity (black curve) and temperature (blue curve) over time. Key phases include I (initial phase), where starch granules begin to hydrate; II (pasting onset), characterized by a rapid viscosity increase due to gelatinization; III (peak viscosity, holding phase, and breakdown), where granules reach maximum swelling, followed by viscosity reduction due to granule rupture; and IV (recrystallization), where retrogradation occurs during cooling. This profile highlights the high swelling capacity and retrogradation tendency of native potato starch

to the starch granules' maximum water retention capacity. The height of the peak reflects the starch's swelling and stability before rupture.

Following the peak, the graph shows a downward slope as granule rupture occurs due to prolonged heating and mechanical shear, resulting in a decrease in viscosity (III). This phase, known as breakdown, reflects the structural instability of the granules. The line then stabilizes into a plateau during the holding phase at high temperatures, representing the resistance of the gel to further reduction in viscosity. As the temperature decreases in the cooling phase, the line slopes upward again, reflecting retrogradation as amylose and amylopectin chains re-associate and form a gel structure (IV). The final plateau indicates the final viscosity, representing the paste's gel strength and stability. The difference between the final and trough viscosities, known as setback viscosity, quantifies the extent of retrogradation. The overall shape and transitions of the curve provide comprehensive insights into the thermal and mechanical properties of the starch during gelatinization, breakdown, and retrogradation.

Pasting temperature provides insights into the point at which granule swelling begins, while peak viscosity indicates the maximum swelling capacity of the granules before structural breakdown. Breakdown viscosity, the difference between peak and hold viscosity, highlights the granule's stability under mechanical stress and heat. Potato starch, for instance, exhibits a higher peak viscosity and stability than wheat starch due to its larger granules and higher amylopectin content. Studies reveal that annealing starch prior to gelatinization narrows the temperature range (T_c-T_o) and increases gelatinization enthalpy, likely due to crystallite growth or improved interactions between crystalline and amorphous regions [34].

Starch gelatinization can be effectively studied using polarized light microscopy and standard optical microscopy, which reveal the structural and optical changes granules undergo during heating in water. Polarized light microscopy is particularly useful for observing the crystalline regions of starch, as native granules exhibit birefringence, visible as a "Maltese cross" pattern. This pattern results from the highly ordered arrangement of amylopectin double helices within the crystalline lamellae [35]. At the onset of gelatinization, as water penetrates the amorphous regions and swelling begins, the birefringence starts to fade, especially at the periphery of the granules, indicating the initial disruption of crystalline order [34].

As the temperature increases further, the birefringence disappears completely, marking the melting of amylopectin crystalline regions and the transition of granules to an amorphous state. Under standard optical microscopy, granules appear swollen and increasingly translucent as water absorption progresses, correlating with the breakdown of internal structures [36]. This loss of structural integrity is seen as a change in granule shape and size, with some granules showing surface fissures. The disappearance of birefringence observed in polarized light microscopy coincides with the completion of crystalline melting and represents a key indicator of gelatinization [37].

At the peak of gelatinization, optical microscopy reveals granules that have expanded to their maximum size, often exhibiting irregular shapes or partial rupture. These changes reflect the combined effects of swelling and mechanical weakening of the granule walls. Polarized light microscopy confirms the loss of crystallinity, as the granules no longer exhibit the Maltese cross pattern [35]. Together, these microscopy techniques provide a comprehensive view of the gelatinization process, illustrating the sequential breakdown of starch granules' structural and optical properties under heating and hydration.

Fourier-transform infrared (FTIR) and Raman spectroscopy capture the disruption of hydrogen bonds and realignment of starch chains. These tools reveal that gelatinization is a cooperative process where amorphous regions hydrate and expand first, straining and eventually disrupting the crystalline regions [35]. Together, these methods provide a comprehensive understanding of the gelatinization process.

1.4.3 Factors Affecting Gelatinization

Gelatinization is the irreversible process in which starch granules absorb water, swell, and lose their crystalline structure upon heating. Several intrinsic and extrinsic factors influence the gelatinization process.

1.4.3.1 Amylose-to-Amylopectin Ratio

The amylose and amylopectin content significantly influence gelatinization. Amylopectin, with its branched structure, forms crystalline regions that determine gelatinization temperatures. Therefore, starches with higher amylopectin content typically exhibit higher gelatinization temperatures and enthalpy due to the stability of their crystalline regions [38]. In contrast, starches with higher amylose content facilitate granule swelling, leading to lower gelatinization temperatures. Zhao et al. [39] highlighted that amylose-rich starches gelatinize faster due to the weaker molecular interactions in their crystalline regions.

1.4.3.2 Granule Size and Morphology

Granule size and surface structure affect the hydration and thermal transitions of starch. Larger granules, such as those in potato starch, require more energy to gelatinize due to their compact and dense crystalline regions [34]. Granules with irregular shapes or surface pores allow for faster water penetration, lowering the onset temperature for gelatinization. Lu et al. [40] noted that starch granules with smooth surfaces exhibit slower water absorption than those with surface irregularities.

1.4.3.3 Water Availability

Water is essential for gelatinization, disrupting hydrogen bonds within starch granules. Limited water restricts granule swelling and gelatinization, while excessive water can dilute the starch matrix and lower thermal transition temperatures [38]. Liu et al. [41] found that optimal water-to-starch ratios are crucial in controlling the gelatinization process in various food formulations.

1.4.3.4 Temperature and Heating Rate

The heating ramp impacts the uniformity and extent of gelatinization. Rapid heating can cause uneven swelling and rupturing of granules, while slower heating allows for a more uniform transition. Tako et al. [42] observed that gradual heating enhances the consistency of gelatinization across different starch granules. Zhang et al. [43] further demonstrated that slower heating rates improve gelatinization by reducing granule fragmentation.

1.4.3.5 Presence of Additives

Minor components such as salts, sugars, and proteins modify the gelatinization process by altering water-binding capacity and molecular interactions within the starch matrix. Zhao et al. [39] showed that lipids form complexes with amylose, delaying gelatinization by stabilizing crystalline regions. Additionally, salts like NaCl increase gelatinization temperatures by strengthening hydrogen bonds between water and starch molecules [44].

1.4.3.6 Lipids and Amylose-Lipid Complexes

Lipids interact predominantly with amylose through hydrophobic interactions and hydrogen bonding to form amylose-lipid complexes. These complexes are stabilized by the insertion of the lipid's nonpolar tail into the hydrophobic core of the amylose helix, creating a helical inclusion complex with a hydrophilic exterior. This structure hinders the mobility of amylose chains, reduces amylose leaching, and limits granule swelling during heating [39]. Therefore, lipids increase the gelatinization temperature (To) and enthalpy (ΔH), as additional thermal energy is required to dissociate these complexes and disrupt the crystalline regions of the granule [35]. Wadikar et al. [45] observed that longer and more saturated fatty acid chains enhanced the stability of these complexes, creating steric hindrance and reducing water penetration into the granules. These interactions primarily affect amylose but can extend to amylopectin branch points, stabilizing the granule's structural integrity and limiting swelling under heat [46].

1.4.3.7 Sugars and Salts

Sugars interact with starch molecules by competing for water through hydrogen bonding. Their hygroscopic nature binds free water molecules, reducing the availability of water for hydration of starch granules. This interaction primarily affects the amorphous regions of amylopectin and delays the swelling of granules, raising the gelatinization temperature [47]. Sugars also stabilize amylopectin helices by forming hydrogen bonds with hydroxyl groups on amylopectin chains, which reinforces crystalline lamellae and further delays the transition to an amorphous state [48].

Salts influence starch gelatinization through ionic interactions and hydrogen bond stabilization. Sodium chloride, for instance, strengthens hydrogen bonds between water and hydroxyl groups in amylopectin and amylose, increasing the energy required to break these bonds [34, 44]. In contrast, certain salts like potassium chloride reduce gelatinization enthalpy by screening electrostatic repulsions within starch polymers, facilitating water penetration and crystalline disruption [49]. These ionic interactions can alter the hydration shell around starch molecules, modulating the granule's thermal behavior and structural stability.

1.4.3.8 Hydrocolloids and Cellulose Nanocrystals

Hydrocolloids, such as guar gum and xanthan gum, create a viscous matrix around starch granules, physically restricting water mobility and delaying hydration. These polysaccharides interact with starch chains via hydrogen bonding, particularly at hydroxyl-rich sites on amylose and amylopectin molecules [50]. This interaction reinforces the granule surface, increasing gelatinization temperature and reducing amylose leaching.

Cellulose nanocrystals act as nano-scale reinforcements, binding to amylose and amylopectin through hydrogen bonding and van der Waals forces. These interactions restrict the mobility of starch chains and create a rigid network within the granule matrix. Chipón et al. [36] demonstrated that cellulose nanocrystals reduce peak viscosity by stabilizing the granule structure and limiting swelling, delaying the disintegration of crystalline and amorphous regions during heating.

1.4.3.9 Proteins and Enzymes

Proteins associated with starch granules contribute to structural stabilization by forming hydrogen bonds with amylose and amylopectin. These bonds restrict water penetration and granule swelling, raising gelatinization temperatures [42]. For example, endogenous proteins such as gluten form a network with starch molecules, enhancing granule rigidity and delaying gelatinization onset.

Enzymes, however, can either stabilize or disrupt starch granules. Transglutaminase catalyzes covalent cross-linking between glutamine and lysine residues on granule-associated proteins, enhancing granule rigidity and reducing swelling power [51]. In contrast, proteases hydrolyze these protein bonds, exposing internal starch chains to water, facilitating granule hydration, and lowering gelatinization temperatures. These enzymatic treatments alter the starch's thermal behavior by selectively modifying the molecular environment within granules.

1.4.3.10 Mechanical and Chemical Modifications

Mechanical treatments, such as ball milling, disrupt the crystalline regions of starch granules, exposing amylose and amylopectin chains to water. Oliveira Barros et al. [52] reported that these disruptions reduce gelatinization temperature and promote earlier hydration. When combined with salts, mechanical treatments amplify ionic interactions, stabilizing water-starch hydrogen bonds or facilitating the penetration of ions into granules, depending on the salt concentration.

Chemical cross-linking increases granule rigidity by forming covalent bonds between amylose and amylopectin chains, stabilizing both crystalline and amorphous regions. Lian et al. [53] demonstrated that cross-linked starch granules exhibit narrowed gelatinization temperature ranges and increased ΔH, reflecting enhanced structural stability under thermal stress. These treatments enable precise control over starch gelatinization, tailoring its properties for specific applications.

Gamma irradiation and other physical treatments also interact with additives to influence gelatinization. Peng et al. [54] demonstrated that gamma irradiation alters the amylose and amylopectin chain length, making them more susceptible to additive interactions. For example, irradiated starch granules exhibited higher gelatinization temperatures in the presence of salts, suggesting that structural modifications enhance the ionic effects of salts on starch-water interactions.

1.4.3.11 Dissociation During Gelatinization

Retrogradation begins with gelatinization, a phase where starch granules absorb water, swell, and lose their native semi-crystalline structure due to the disruption of hydrogen bonds by heat and moisture. This process allows water to penetrate the granules, breaking hydrogen bonds and hydrating amylose and amylopectin chains. The granules swell, lose their birefringence, and release these polysaccharides into the surrounding medium, forming a viscous and amorphous gel [34, 55]. During gelatinization, the granular architecture dissolves, enabling linear amylose chains and branched amylopectin molecules to disperse fully. This dissociation increases molecular mobility and exposes reactive hydroxyl groups, setting the stage for recrystallization during cooling.

1.4.3.12 Rapid Amylose Reassociation (Short-Term Retrogradation)

Upon cooling, amylose chains rapidly realign into double helices due to their linear structure, driven by hydrogen bonding and van der Waals interactions. This short-term retrogradation typically occurs within hours and results in the formation of crystalline microdomains within the gel network, providing initial firmness and stabilizing the structure [40, 42]. This phase is essential for products like sauces and gels [56].

The crystalline domains formed during this phase can be classified into two main types:

- **B-type crystals**: Hydrated hexagonal arrangements of double helices stabilized by water molecules within interstitial spaces.
- **V-type crystals**: Single helices formed through amylose complexation with lipids or other small molecules, stabilized by hydrophobic and hydrogen bonding interactions.

The cooling rate plays a significant role in the crystallization pattern. Slow cooling promotes the formation of larger, more stable crystalline regions, while rapid cooling results in smaller, less ordered domains [34]. These microcrystals serve as a scaffold for further structural stabilization and enhance the gel's resistance to deformation.

Physically, this phase transforms the gel from an amorphous, fluid-like state into a firmer, semi-solid network. The rapid reassociation of amylose chains reduces molecular mobility and decreases the water-binding capacity of the system, further solidifying the gel matrix.

1.4.3.13 Amylopectin Crystallization (Long-Term Retrogradation)

Amylopectin, with its highly branched structure, retrogrades more slowly than amylose. Over weeks or months, its outer branches gradually form double helices that align into semi-crystalline regions, a process requiring extensive molecular adjustments. This slower, yet more thermodynamically stable crystallization results in the formation of distinct crystalline structures that contribute to the gel's long-term stability [53, 57]. The slower recrystallization of amylopectin enhances gel rigidity [49].

Two primary types of crystalline structures are associated with amylopectin retrogradation:

- **A-type crystals**: Compact, dense structures with minimal water content, commonly found in amylopectin-rich starches.
- **C-type crystals**: Hybrid structures that combine features of A- and B-type polymorphs, typically forming under intermediate moisture conditions.

These semi-crystalline regions enhance the gel's opacity and firmness, reducing its susceptibility to deformation. The degree of crystallinity, detectable through X-ray diffraction, is influenced by factors such as amylopectin chain length, temperature, and moisture content [54]. Longer amylopectin chains contribute to sharper diffraction peaks, indicating highly ordered crystalline regions.

This phase is critical for the gel's long-term stability, as it transitions into a denser and more rigid structure. Physically, amylopectin crystallization leads to hardening in products like bread, a phenomenon known as staling, or increased rigidity in bioplastic films. These structural transformations ensure stability over extended periods while affecting the material's texture and mechanical properties.

1.4.3.14 Crystalline Stabilization

As retrogradation progresses, the crystalline regions of amylose and amylopectin mature and stabilize, reinforcing the gel's structural integrity. Molecular interactions, such as hydrogen bonding and hydrophobic effects, play a key role in this stabilization process. The resulting crystalline structures exhibit characteristic B-type or V-type patterns, as identified in X-ray diffraction studies [40, 53]. This stabilization reduces molecular mobility within the gel matrix, minimizing syneresis (water exudation) and enhancing the material's resistance to mechanical and thermal stresses.

Physically, the gel becomes less permeable to water and develops greater structural resilience. Advanced techniques, such as small-angle X-ray scattering and scanning electron microscopy, have revealed thickening of lamellar layers and the formation of ordered microstructures that define the final properties of retrograded starch [55].

Prolonged retrogradation can give rise to unique structural patterns beyond conventional crystalline polymorphs, including:

- **Intercalated structures**: Complexes formed by amylose or amylopectin chains interacting with salts or polar molecules, producing hybrid crystalline forms with distinctive diffraction patterns.
- **Layered assemblies**: Alternating crystalline and amorphous layers that reflect the molecular periodicity and organization within retrograded starch systems [55].

The stabilization of these structures is governed by hydrogen bonding, van der Waals forces, and hydrophobic interactions. This molecular organization enhances the gel's thermal resistance and defines its functional properties, making retrograded starch suitable for various applications in food, packaging, and biodegradable materials. By minimizing syneresis and improving structural stability, these advanced crystalline forms contribute to the durability and performance of starch-based systems.

1.5 Structural Characterization Techniques

1.5.1 Differential *Scanning Calorimetry (DSC)*

DSC is one of the most common thermal analysis techniques used to study the retrogradation process. It measures the enthalpy change (ΔH) associated with the melting of retrograded starch, reflecting the extent of crystallization [38, 40]. By comparing the ΔH values of fresh and stored samples, researchers can calculate the retrogradation degree (RD). The higher the ΔH, the more crystalline regions have formed.

For example, Donmez et al. [38] found that starches with higher amylose content exhibit greater retrogradation enthalpy, indicating more significant recrystallization. Moreover, DSC thermograms display distinct endothermic peaks corresponding to the melting temperatures of retrograded amylose and amylopectin crystals, providing insights into the stability of these crystalline regions over time.

1.5.2 Small-Angle X-Ray Scattering (SAXS)

SAXS provides nanoscale insights into the structural organization of retrograded starch. This technique measures the radius of gyration and lamellar repeat lengths, revealing periodic structures formed during retrogradation [40].

SAXS patterns indicate that amylose and amylopectin chains aggregate into lamellar structures, with increasing thickness over time. This progressive ordering signifies the formation of stable crystalline regions. The technique is particularly useful for examining the development of amylopectin crystallites, which contribute to long-term retrogradation.

1.5.3 X-Ray Diffraction (XRD)

XRD is a critical tool for assessing the crystalline patterns in retrograded starch. Retrogradation typically results in a transition from amorphous to more ordered crystalline structures, with amylose forming V-type crystals and amylopectin forming B-type crystals [53].

The intensity and sharpness of diffraction peaks in XRD patterns indicate the extent of crystallinity. Lu et al. [40] observed that starch stored at lower temperatures exhibits sharper XRD peaks, indicating more extensive crystallization. XRD analysis also provides information on polymorphic transitions, which are crucial for understanding the stability and functionality of retrograded starch.

1.5.4 Scanning Electron Microscopy (SEM)

SEM is used to visualize the microstructural changes in starch granules during retrogradation. It reveals surface alterations, such as the development of crystalline clusters and cracks [40, 42].

SEM images of retrograded starch granules show substructures composed of smaller primary particles, suggesting that amylose and amylopectin chains rearrange into distinct clusters. This technique provides direct evidence of the morphological transformations occurring during retrogradation.

1.5.5 *High-Performance Anion-Exchange Chromatography (HPAEC)*

HPAEC is employed to analyze the chain length distribution of starch molecules in retrograded samples. The distribution of longer chains correlates with more stable crystalline regions, influencing the retrogradation process [55].

Niu et al. [55] found that retrograded starch contains chains of various lengths, with longer chains contributing to the formation of semi-crystalline structures. This distribution affects the particle size and stability of retrograded starch, highlighting the importance of molecular composition in retrogradation.

1.5.6 *Microscopy with Deep Learning*

A novel approach developed by Niu et al. [55] combines microscopy with deep learning to analyze starch retrogradation. This method utilizes computer vision techniques to track changes in starch granule morphology and birefringence under polarized light.

The deep learning algorithm enhances the accuracy of quantifying the retrogradation process, providing high-throughput data on starch gel morphology. This innovative approach allows researchers to gain detailed insights into the dynamics of retrogradation at the microscopic level.

1.6 Starch Retrogradation

Retrogradation is a post-gelatinization process where starch molecules, primarily amylose and amylopectin, reassociate into ordered structures during cooling and storage. This recrystallization process profoundly influences the texture, stability, and shelf life of starch-based materials. It is governed by molecular rearrangements and physical transformations, which occur sequentially and are influenced by environmental factors and starch composition. Retrogradation involves a sequence of molecular events, starting from the disruption of gelatinized starch to the gradual formation of double helices and crystalline aggregates.

1.6.1 *Factors Affecting Retrogradation*

Retrogradation is the process where gelatinized starch chains realign into more ordered structures during cooling and storage. Several factors influence the rate and extent of retrogradation:

1.6.1.1 Amylose and Amylopectin Structure

Amylose plays a dominant role in the early stages of retrogradation due to its linear structure, which allows rapid hydrogen bonding and crystallization upon cooling. Amylopectin, with its branched chains, retrogrades more slowly, contributing to the long-term stability of the gel [53]. Zhao et al. [39] noted that longer amylopectin chains form more stable crystalline regions over time, enhancing the mechanical stability of retrograded starch gels.

1.6.1.2 Storage Temperature and Time

Low temperatures accelerate amylose retrogradation by promoting hydrogen bonding and crystalline structure formation. Donmez et al. [38] highlighted that refrigeration storage leads to faster retrogradation in the initial phase, while amylopectin crystallization occurs gradually over weeks. Gao et al. [58] demonstrated that prolonged storage enhances retrogradation, particularly for amylopectin, as its slower crystallization process contributes to gel stability over time.

1.6.1.3 Moisture Content

Moisture availability significantly impacts retrogradation. Higher moisture levels increase molecular mobility, facilitating chain realignment and crystalline formation. Conversely, low moisture content restricts mobility and reduces the extent of retrogradation [34]. Niu et al. [55] emphasized that intermediate moisture levels optimize retrogradation, improving gel firmness and stability.

1.6.1.4 Crystallinity and Structural Rearrangements

The degree of crystallinity in retrograded starch influences its mechanical and thermal stability. Lian et al. [53] noted that retrograded starch exhibits B-type crystalline patterns, which are more stable and less prone to enzymatic hydrolysis. X-ray diffraction (XRD) analysis shows that the formation of these crystalline structures depends on the molecular composition and storage conditions [54]. Jane et al. [59] highlighted that higher crystallinity levels reduce water-binding capacity, impacting the textural properties of retrograded starch products.

1.6.1.5 Additives and Treatments

Additives, including lipids, proteins, and salts, modify retrogradation by influencing molecular interactions and water redistribution. Zhao et al. [39] observed that amylose-lipid complexes reduce the availability of amylose for recrystallization during retrogradation. Dang et al. [51] showed that transglutaminase enzymes stabilize the starch structure by forming cross-links, reducing retrogradation, while proteases enhance retrogradation by increasing water accessibility.

Starch retrogradation is influenced by various additives that interact with starch molecules, altering their recrystallization behavior. These additives, including lipids, proteins, carbohydrates, salts, polyphenols, organic acids, and emulsifiers, modify the rate and extent of retrogradation by impacting molecular interactions such as hydrogen bonding, water mobility, and structural stabilization. Understanding these interactions is crucial for controlling the textural and functional properties of starch-based products, thereby extending their shelf life and improving their quality. This document explores the impact of these additives on starch retrogradation using a range of references from recent studies.

Lipids are among the most studied additives that affect starch retrogradation. During gelatinization, amylose chains can form complexes with lipids, creating amylose-lipid complexes that inhibit retrogradation. These complexes reduce the mobility of amylose molecules, thereby limiting their ability to recrystallize. The strength and stability of the amylose-lipid complex depend on factors such as the chain length of the fatty acid and the processing conditions. For instance, long-chain fatty acids form more stable complexes that effectively reduce amylose retrogradation [57].

Zhao et al. [39] demonstrated that the presence of fatty acids or phospholipids increases the thermal stability of gelatinized starch by preventing the recrystallization of amylose during storage. Additionally, lipids can interfere with the interaction between amylopectin molecules, further reducing the formation of crystalline structures. Baranowska et al. [46] observed that the incorporation of lipids decreased the crystallinity of retrograded starch, as confirmed by a reduction in X-ray diffraction (XRD) peak intensity. This effect delays retrogradation and improves the shelf life of starch-based foods by maintaining their softness and reducing syneresis.

Proteins, both endogenous and added, significantly influence the retrogradation process by interacting with starch molecules through hydrogen bonding and hydrophobic interactions. For example, whey protein isolate and soy protein isolate (SPI) have been shown to retard retrogradation by reducing the formation of amylose and amylopectin crystallites [57]. The presence of proteins dilutes the starch network and reduces water availability, leading to a softer gel structure during storage [40].

Jimenez et al. [60] found that caseinate reduced the crystallinity of starch-based films, suggesting that proteins hinder the rearrangement of starch chains into ordered structures. Proteins can also form cross-links with starch molecules, affecting the formation of double helices and reducing the degree of retrogradation. Wang et al. [49] highlighted those proteins form hydrogen bonds with starch polymers, interfering with amylose and amylopectin reassociation, leading to softer gels and reduced syneresis.

Carbohydrates, such as monosaccharides, disaccharides, oligosaccharides, and polysaccharides, have mixed effects on retrogradation. Some studies suggest that polysaccharides can slow or prevent retrogradation by interfering with the alignment of starch chains and reducing water availability. Donmez et al. [38]

highlighted that sucrose delays retrogradation by interfering with the reassociation of amylopectin chains into double helices.

Polysaccharides such as guar gum interact with starch molecules through hydrogen bonding, modifying the water distribution in the system and thereby affecting the retrogradation process [55]. Chipón et al. [36] demonstrated that the addition of cellulose nanocrystals to starch gels reinforced the gel matrix, inhibiting retrogradation by creating physical barriers that impede the alignment of starch molecules into crystalline regions.

Salts affect starch retrogradation by modifying the ionic interactions within the starch-water system. The type, valency, and concentration of salts can influence the retrogradation rate and the thermal and rheological properties of starch [57]. Bivalent cations such as calcium (Ca^{2+}) and magnesium (Mg^{2+}) are more effective in reducing the retrogradation rate compared to monovalent cations such as sodium (Na^{+}) and potassium (K^{+}) due to their stronger interaction with starch hydroxyl groups.

Beck et al. [61] demonstrated that salts reduce the mobility of amylose molecules by replacing hydrogen ions with cations, impacting the hydrogen bonding within the starch gel. Enpeng Li et al. [44] found that NaCl enhances the retrogradation rate by stabilizing the interactions between starch polymers and water molecules, leading to faster amylose reassociation into crystalline structures.

Polyphenols and organic acids, such as citric acid, also affect retrogradation by modifying the pH of the starch system and interacting with starch molecules through hydrogen bonding and hydrophobic interactions. These interactions can hinder the reassociation of amylose and amylopectin chains, reducing the formation of crystalline regions [57].

Polyphenols may also act as antioxidants, protecting starch molecules from oxidative damage, which can further impact retrogradation. Lu et al. [40] noted that organic acids change the solubility of starch molecules and reduce the availability of water for crystallization, thereby slowing the retrogradation process.

Finally, Emulsifiers such as monoglycerides and lecithin interact with amylose chains to form inclusion complexes that reduce retrogradation. These complexes inhibit the crystallization of amylose and reduce the firmness of starch gels during storage. Emulsifiers are commonly used in bakery products to improve texture and extend shelf life by reducing staling [57].

Jagannadham et al. [56] found that emulsifiers reduce amylose leaching and inhibit crystallization, thereby minimizing gel hardening over time. This stabilization improves the textural quality and storage stability of starch-based products.

1.7 Conclusions

A Starch is a fundamental biopolymer with diverse applications in food, pharmaceuticals, and biodegradable materials. Its functionality is primarily determined by its structural composition, specifically the ratio of amylose to amylopectin, which influences gelatinization, retrogradation, and crystallinity. Advancements in starch modification techniques—such as enzymatic, chemical, and physical treatments—have enhanced its thermal stability, mechanical properties, and resistance to retrogradation, broadening its industrial utility. Additionally, research on non-conventional starch sources, including plantain and achira, has unveiled new opportunities for tailored applications in sustainable packaging, food texturization, and biopolymers.

Further exploration of underutilized starch sources is recommended to diversify industrial applications and improve sustainability. Expanding research on these alternative starches can enhance the development of eco-friendly materials, improve food formulations, and reduce reliance on conventional starch sources, contributing to a more sustainable and resource-efficient industry. Understanding the interactions between starch structure and function is essential for optimizing its performance across various industries. Ongoing research will continue to expand its potential, addressing sustainability challenges while improving product quality and functionality.

Acknowledgments The authors would like to thank to the Programa Iberoamericano de Ciencia y Tecnología para el Desarrollo (CYTED) for its partial support of this work through Red 121RT0108. Additionally, L. Ávila-Martín acknowledges the financial support received from Becas Bicentenario Corte 1, provided by the Ministerio de Ciencia, Tecnología e Innovación, Colombia.

Conflicts of Interest The authors declare no conflict of interest.

References

1. Sudheesh C, Varsha L, Sunooj KV, Pillai S. Influence of crystalline properties on starch functionalization from the perspective of various physical modifications: a review. Int J Biol Macromol. 2024;280:136059.
2. Shi Q, Xia Y, Xue N, Wang Q, Tao Q, Li M, et al. Modulation of starch synthesis in Arabidopsis via phytochrome B-mediated light signal transduction. J Integr Plant Biol [Internet]. 2024;66(5):973–85. [cited 2025 Feb 4]. Available from: https://onlinelibrary.wiley.com/doi/full/10.1111/jipb.13630.
3. Ahmad D, Ying Y, Bao J. Understanding starch biosynthesis in potatoes for metabolic engineering to improve starch quality: a detailed review. Carbohydr Polym. 2024;346:122592.
4. Bangar SP, Whiteside WS, Chowdhury A, Ilyas RA, Siroha AK. Recent advancements in functionality, properties, and applications of starch modification with stearic acid: a review. Int J Biol Macromol. 2024;280:135782.
5. Carvalho HJM, Barcia MT, Schmiele M. Non-conventional starches: properties and potential applications in food and non-food products. Macromolecules. 2024;4:886–909. [Internet]. 2024 Dec 17 [cited 2025 Feb 4];4(4):886–909. Available from: https://www.mdpi.com/2673-6209/4/4/52/htm.
6. Dorantes-Fuertes MG, López-Méndez MC, Martínez-Castellanos G, Meléndez-Armenta RÁ, Jiménez-Martínez HE. Starch extraction methods in tubers and roots: a systematic review. Agronomy. 2024;14:865. [Internet]. 2024 Apr 20 [cited 2025 Feb 5];14(4):865. Available from: https://www.mdpi.com/2073-4395/14/4/865/htm.

7. Ferreiro O, Martin LA, Chacon WDC, Duarte S, Perilla JE, Ayala Valencia G. Valorization of agro-industrial plantain (Musa × paradisiaca) by-products: alternative sources of carbohydrates and bioactive compounds. Starch - Stärke [Internet]. 2023;76:2300210. [cited 2024 Jan 23]. Available from: https://onlinelibrary.wiley.com/doi/full/10.1002/star.202300210.
8. Ávila-Martín L, Beltrán-Osuna ÁA, Perilla JE. Effect of the addition of citric acid and whey protein isolate in Canna indica L. starch films obtained by solvent casting. J Polym Environ [Internet]. 2020;28(3):871–83. [cited 2024 Jun 28]. Available from: https://link.springer.com/article/10.1007/s10924-019-01648-z.
9. Seeponkai N, Pundi S, Fuggate P, Kanla S, Khammuang K, Seephonkai P. StarchCrete: a starch-based biocomposite concrete for lightweight building material applications. Ind Crop Prod. 2024;222:119991.
10. Hizukuri S. Polymodal distribution of the chain lengths of amylopectins, and its significance. Carbohydr Res. 1986;147(2):342–7.
11. Kong X. Molecular structure of starch. In: Physical modifications of starch [Internet]; 2023. p. 1–11. [cited 2024 Dec 12]. Available from: https://link.springer.com/chapter/10.1007/978-981-99-5390-5_1.
12. Zhong Y, Tai L, Blennow A, Ding L, Herburger K, Qu J, et al. High-amylose starch: structure, functionality and applications. Crit Rev Food Sci Nutr [Internet]. 2023;63(27):8568–90. [cited 2024 Dec 12]. Available from: https://www.tandfonline.com/doi/abs/10.1080/10408398.2022.2056871.
13. Pérez S, Bertoft E. The molecular structures of starch components and their contribution to the architecture of starch granules: a comprehensive review. Starch - Stärke [Internet]. 2010;62(8):389–420. [cited 2022 Feb 22]. Available from: https://onlinelibrary.wiley.com/doi/full/10.1002/star.201000013.
14. Tester RF, Karkalas J, Qi X. Starch—composition, fine structure and architecture. J Cereal Sci. 2004;39(2):151–65.
15. Li H, Wen Y, Wang J, Sun B. Relations between chain-length distribution, molecular size, and amylose content of rice starches. Int J Biol Macromol. 2018;120:2017–25.
16. Buléon A, Colonna P, Planchot V, Ball S. Starch granules: structure and biosynthesis. Int J Biol Macromol. 1998;23(2):85–112.
17. Radosta S, Haberer M, Vorwerg W. Molecular characteristics of amylose and starch in dimethyl sulfoxide. Biomacromolecules [Internet]. 2001;2(3):970–8. [cited 2024 Apr 15]. Available from: https://pubs.acs.org/doi/abs/10.1021/bm0100662.
18. Agama-Acevedo E, Nuñez-Santiago MC, Alvarez-Ramirez J, Bello-Pérez LA. Physicochemical, digestibility and structural characteristics of starch isolated from banana cultivars. Carbohydr Polym. 2015;124:17–24.
19. Trinh KS, Dang TB. Structural, physicochemical, and functional properties of electrolyzed cassava starch. Int J Food Sci [Internet]. 2019;2019. [cited 2024 Apr 15]. Available from: /pmc/articles/PMC6525864/.
20. Zhu F. Barley starch: composition, structure, properties, and modifications. Compr Rev Food Sci Food Saf [Internet]. 2017;16(4):558–79. [cited 2024 Apr 12]. Available from: https://onlinelibrary.wiley.com/doi/full/10.1111/1541-4337.12265.
21. Tang H, Mitsunaga T, Kawamura Y. Molecular arrangement in blocklets and starch granule architecture. Carbohydr Polym. 2006;63(4):555–60.
22. Chávez-Salazar A, Álvarez-Barreto CI, Hoyos-Leyva JD, Bello-Pérez LA, Castellanos-Galeano FJ. Drying processes of OSA-modified plantain starch trigger changes in its functional properties and digestibility. LWT. 2022;154:112846.
23. Daza LD, Montealegre MÁ, Reche C, Sandoval-Aldana A, Eim VS, Váquiro HA. Chachafruto starch: physicochemical characterization, film-forming properties, and 3D printability. Int J Biol Macromol. 2023;247:125795.
24. De Dios-Avila N, Tirado-Gallegos JM, Rios-Velasco C, Luna-Esquivel G, Isiordia-Aquino N, Zamudio-Flores PB, et al. Physicochemical, structural, thermal and rheological properties of flour and starch isolated from avocado seeds of landrace and Hass cultivars. Molecules

[Internet]. 2022;27(3):910. [cited 2022 Feb 12]. Available from: https://www.mdpi.com/1420-3049/27/3/910.
25. Wei M, Andersson R, Xie G, Salehi S, Boström D, Xiong S. Properties of cassava stem starch being a new starch resource. Starch - Stärke [Internet]. 2018;70(5–6):1700125. [cited 2024 Apr 10]. Available from: https://onlinelibrary.wiley.com/doi/full/10.1002/star.201700125.
26. Spinozzi F, Ferrero C, Perez S. The architecture of starch blocklets follows phyllotaxic rules. Sci Rep. 2020;10:1. [Internet]. 2020 Nov 18 [cited 2024 Dec 12];10(1):1–16. Available from: https://www.nature.com/articles/s41598-020-72218-w.
27. Bertoft E. Understanding starch structure: recent progress. Agronomy. 2017;7:56. [Internet]. 2017 Aug 25 [cited 2024 Dec 12];7(3):56. Available from: https://www.mdpi.com/2073-4395/7/3/56/htm.
28. Barandiaran A, Montanes N, Gomez-Caturla J, Balart R, Florez-Prieto MA, Ávila-Martin L, et al. Development and characterization of edible films based on starch isolated from different Colombian potato varieties. Int J Biol Macromol. 2024;263:130165.
29. Cai C, Zhao L, Huang J, Chen Y, Wei C. Morphology, structure and gelatinization properties of heterogeneous starch granules from high-amylose maize. Carbohydr Polym. 2014;102(1):606–14.
30. Guo L, Chen H, Zhang Y, Yan S, Chen X, Gao X. Starch granules and their size distribution in wheat: biosynthesis, physicochemical properties and their effect on flour-based food systems. Comput Struct Biotechnol J. 2023;21:4172–86.
31. Xu M, Xu C, Kim SJ, Ji S, Ren Y, Chen Z, et al. Investigating the evolution of the fine structure in cassava starch during growth and its correlation with gelatinization performance. Int J Biol Macromol. 2024;265:130422.
32. Wang H, Feng Y, Guo K, Shi L, Xu X, Wei C. Structural, thermal, pasting and digestion properties of starches from developing root tubers of sweet potato. Food Secur. 2024;13:1103. [Internet]. 2024 Apr 3 [cited 2025 Jan 20];13(7):1103. Available from: https://www.mdpi.com/2304-8158/13/7/1103/htm.
33. Yang L, Li L, Jiang F, Zhang Q, Yang F, Wang Y, et al. Structural and physicochemical characteristics of starches from sorghum varieties with varying amylose content. Food Sci Nutr [Internet]. 2024;12(10):7989–99. [cited 2025 Jan 20]. Available from: https://onlinelibrary.wiley.com/doi/full/10.1002/fsn3.4245.
34. Jacobs H, Eerlingen RC, Delcour JA. Factors affecting the visco-amylograph and rapid visco-analyzer evaluation of the impact of annealing on starch pasting properties. Starch - Stärke [Internet]. 1996;48(7–8):266–70. [cited 2024 Dec 25]. Available from: https://onlinelibrary.wiley.com/doi/full/10.1002/star.19960480707.
35. Chakraborty I, N P, Mal SS, Paul UC, Rahman MH, Mazumder N. An insight into the gelatinization properties influencing the modified starches used in food industry: a review. Food Bioprocess Technol. 2022;15:6. [Internet]. 2022 Feb 3 [cited 2024 Dec 25];15(6):1195–223. Available from: https://link.springer.com/article/10.1007/s11947-022-02761-z.
36. Chipón J, Ramírez K, Morales J, Díaz-calderón P. Rheological and thermal study about the gelatinization of different starches (potato, wheat and waxy) in blend with cellulose nanocrystals. Polymers (Basel) [Internet]. 2022;14(8):1560. [cited 2024 Dec 25]. Available from: https://pubmed.ncbi.nlm.nih.gov/35458308/.
37. Huang S, Chao C, Yu J, Copeland L, Wang S. New insight into starch retrogradation: the effect of short-range molecular order in gelatinized starch. Food Hydrocoll. 2021;120:106921.
38. Donmez D, Pinho L, Patel B, Desam P, Campanella OH. Characterization of starch–water interactions and their effects on two key functional properties: starch gelatinization and retrogradation. Curr Opin Food Sci. 2021;39:103–9.
39. Zhao T, Zhang H, Chen F, Tong P, Cao W, Jiang Y. Study on structural changes of starches with different amylose content during gelatinization process. Starch - Stärke [Internet]. 2022;74(7–8):2100269. [cited 2025 Jan 3]. Available from: https://onlinelibrary.wiley.com/doi/full/10.1002/star.202100269.

40. Lu H, Zhan J, Shen W, Ma R, Tian Y. Assessing starch retrogradation from the perspective of particle order. Food Secur [Internet]. 2024;13:911. 2024 Mar 17 [cited 2024 Dec 27];13(6):911. Available from: https://www.mdpi.com/2304-8158/13/6/911/htm.
41. Liu X, Chao C, Yu J, Copeland L, Wang S. Mechanistic studies of starch retrogradation and its effects on starch gel properties. Food Hydrocoll. 2021;120:106914.
42. Tako M, Tamaki Y, Teruya T, Takeda Y, Tako M, Tamaki Y, et al. The principles of starch gelatinization and retrogradation. Food Nutr Sci [Internet]. 2014;5(3):280–91. [cited 2024 Dec 27]. Available from: https://www.scirp.org/journal/paperinformation?paperid=42262.
43. Zhang L, Zhao J, Li F, Jiao X, Zhang Y, Yang B, et al. Insight to starch retrogradation through fine structure models: a review. Int J Biol Macromol [Internet]. 2024;273:132765. [cited 2024 Jun 25]. Available from: https://linkinghub.elsevier.com/retrieve/pii/S0141813024035700.
44. Li E, Lv J, Huo D, Jia B, Li C. Importance of amylose chain-length distribution in determining starch gelatinization and retrogradation property of wheat flour in the presence of different salts. Carbohydr Polym. 2023;308:120648.
45. Wadikar DD, Khan MA, Chiravi K, Mahesh C, Semwal AD. Starch retrogradation: role of food ingredients and extrinsic factors. Food Sci Eng. 2023;4(2):204–15.
46. Baranowska HM, Sikora M, Krystyjan M, Dobosz A, Tomasik P, Walkowiak K, et al. Analysis of the retrogradation processes in potato starches blended with non-starchy polysaccharide hydrocolloids by LF NMR. Food Biophys [Internet]. 2020;15(1):64–71. [cited 2024 Dec 27]. Available from: https://link.springer.com/article/10.1007/s11483-019-09600-3.
47. Ai Y, Jane JL. Gelatinization and rheological properties of starch. Starch - Stärke [Internet]. 2015;67(3–4):213–24. [cited 2024 Dec 21]. Available from: https://onlinelibrary.wiley.com/doi/full/10.1002/star.201400201.
48. Yao Y, Zhang J, Ding X. Structure–retrogradation relationship of rice starch in purified starches and cooked rice grains: a statistical investigation. J Agric Food Chem [Internet]. 2002;50(25):7420–5. [cited 2024 Dec 27]. Available from: https://pubs.acs.org/doi/abs/10.1021/jf020643t.
49. Wang S, Li C, Copeland L, Niu Q, Wang S. Starch retrogradation: a comprehensive review. Compr Rev Food Sci Food Saf. 2015;14(5):568–85.
50. Lu H, Ma R, Chang R, Tian Y. Evaluation of starch retrogradation by infrared spectroscopy. Food Hydrocoll. 2021;120:106975.
51. Dang LTK, Imaizumi T, Nishizu T. Transglutaminase and protease: how starch granule-associated protein modification affects starch retrogradation. Food Hydrocoll. 2025;158:110551.
52. de Oliveira BM, Mattos ALA, de Almeida JS, de Freitas RM, de Brito ES. Effect of ball-milling on starch crystalline structure, gelatinization temperature, and rheological properties: towards enhanced utilization in thermosensitive systems. Food Secur [Internet]. 2023;12(15):2924. [cited 2024 Dec 22]. Available from: https://www.mdpi.com/2304-8158/12/15/2924/htm.
53. Lian X, Cheng K, Wang D, Zhu W, Wang X. Analysis of crystals of retrograded starch with sharp X-ray diffraction peaks made by recrystallization of amylose and amylopectin. Int J Food Prop [Internet]. 2017;20:S3224–36. [cited 2024 Dec 27]. Available from: https://www.tandfonline.com/doi/abs/10.1080/10942912.2017.1362433.
54. Peng Z, Wang X, Liu Z, Zhang L, Cheng L, Nia J, et al. Modifying the resistant starch content and the retrogradation characteristics of potato starch through high-dose gamma irradiation. Gels. 2024;10:763. [Internet]. 2024 Nov 24 [cited 2024 Dec 27];10(12):763. Available from: https://www.mdpi.com/2310-2861/10/12/763/htm.
55. Niu Y, Zheng Y, Fu X, Zeng D, Liu H. A novel characterization of starch gelatinization using microscopy observation with deep learning methodology. J Food Eng. 2022;327:111057.
56. Jagannadham K, Parimalavalli R, Surendra Babu A. Effect of triple retrogradation treatment on chickpea resistant starch formation and its characterization. J Food Sci Technol [Internet]. 2017;54(4):901–8. [cited 2024 Dec 27]. Available from: https://pubmed.ncbi.nlm.nih.gov/28303041/.

57. Fu Z, Chen J, Luo SJ, Liu CM, Liu W. Effect of food additives on starch retrogradation: a review. Starch - Stärke [Internet]. 2015;67(1–2):69–78. [cited 2024 Dec 27]. Available from: https://onlinelibrary.wiley.com/doi/full/10.1002/star.201300278.
58. Gao D, Li X, Li F, Luo R, Liao H, Man J. Changes of crystalline structure and physicochemical properties of Pueraria lobata var. thomsonii starch under water deficit. PLoS One [Internet]. 2024;19(7):e0304373. [cited 2024 Dec 22]. Available from: https://pmc.ncbi.nlm.nih.gov/articles/PMC11221752/.
59. Jane J, Chen YY, Lee LF, McPherson AE, Wong KS, Radosavljevic M, et al. Effects of amylopectin branch chain length and amylose content on the gelatinization and pasting properties of starch. Cereal Chem. 1999;76(5):629–37.
60. Jiménez A, Fabra MJ, Talens P, Chiralt A. Effect of sodium caseinate on properties and ageing behaviour of corn starch based films. Food Hydrocoll. 2012;29(2):265–71.
61. Beck M, Jekle M, Becker T. Starch re-crystallization kinetics as a function of various cations. Starch - Stärke [Internet]. 2011;63(12):792–800. [cited 2025 Jan 4]. Available from: https://onlinelibrary.wiley.com/doi/full/10.1002/star.201100071.

Chapter 2
Starch in Aerogels

Raul Remor Dalsasso, Gabriel Coelho Leandro, Raquel da Silva Simão, and Wilson Daniel Caicedo Chacon

2.1 Introduction

Aerogels are interconnected porous structures in which the system's solvent is replaced by air [1]. Starch-based aerogels have garnered significant interest in the last decade as sustainable alternatives in materials science. Their ultra-low density, high surface area, and enhanced biocompatibility make them appealing for various applications, including drug delivery systems, food packaging, and thermal insulation materials [2, 3]. Aerogels can be formed from organic [4, 5], inorganic, hybrid, or composite [6–9] raw materials. Starch, a biopolymer derived from abundant and renewable botanical sources, offers distinct advantages over synthetic aerogels. As a biodegradable and non-toxic material, starch aligns well with the growing industrial focus on environmentally friendly processes, addressing ecological concerns associated with synthetic polymers. This emphasis on sustainable materials is increasingly essential, particularly as global awareness of environmental issues rises. Consequently, the ongoing research and development of starch-based aerogels align with global efforts to minimize environmental degradation and promote sustainable materials [3, 10].

The production of starch-based aerogels involves a multistep process designed to achieve and preserve a highly porous network structure, which is crucial to their functionality [11]. The first step, gelatinization, involves heating starch in water,

R. R. Dalsasso (✉) · G. C. Leandro · R. da Silva Simão
Department of Chemical and Food Engineering, Federal University of Santa Catarina, Florianópolis, SC, Brazil
e-mail: raul.remor.dalsasso@posgrad.ufsc.br

W. D. C. Chacon
Department of Chemical and Food Engineering, Federal University of Santa Catarina, Florianópolis, SC, Brazil

Ingeniería de Alimentos, Fundación Universitaria Agraria de Colombia, Bogotá, Colombia

G. Ayala Valencia (ed.), *Starch: Progress in Food Applications*,
https://doi.org/10.1007/978-3-031-98732-8_2

where the granules' crystalline structure is disrupted. This heating allows starch molecules, primarily amylose, and amylopectin, to transition into an amorphous state [11–13]. This transition is vital as it facilitates the formation of a viscous gel, which serves as a precursor to the aerogel. The gelatinization process can vary depending on the type of starch and the conditions used, such as temperature and time. This variability can impact the texture and properties of the resulting gel. Following gelatinization, a retrogradation occurs, wherein the amylose chains realign into a semi-crystalline structure, forming a stable gel matrix [14, 15]. Retrogradation is influenced by cooling rates and the presence of additives, which can enhance the properties of the gel. These steps are critical as they determine the mechanical and thermal properties of the resulting aerogel [16]. Different drying methods can convert the hydrogel to the aerogel by exchanging the water between the polymer chains with air. The most common techniques used are freeze-drying [17–19], vacuum drying [13, 15], and supercritical CO_2 drying [1, 20, 21]. Based on these drying methods, the final material is classified as cryogel, xerogel, or aerogel, respectively. Concerning the pore size and microstructure, aerogels are classified into distinct types, namely microporous (˂2 nm), mesoporous (2–50 nm), and macro-porous (˃50 nm) [22].

The choice of starch type and processing conditions significantly affects the final aerogel's properties. High-amylose starches are preferred when mechanical robustness is required, as they form denser networks that impart strength and resilience. In contrast, amylopectin-rich starches tend to produce weaker gels due to their branched structure, which can hinder the formation of a cohesive network [13]. The amylose-to-amylopectin ratio is thus critical in determining the mechanical performance of the aerogels, influencing their ability to withstand external stresses.

This influence of starch composition on aerogel performance has prompted research into starch blending techniques, where starch is combined with other natural polymers like cellulose or chitosan. For instance, cellulose-starch composites can enhance mechanical stability and hydrophobicity, making the material suitable for packaging that must withstand variable moisture levels. This is particularly relevant in food packaging, where maintaining product integrity against moisture is crucial. Incorporating chitosan, known for its antimicrobial properties, into starch-based aerogels can create biodegradable and functional packaging materials for food preservation [11, 23].

Blending starch with other biopolymers can lead to synergistic effects, enhancing mechanical, thermal, and barrier properties compared to the individual components [24]. Researchers have also explored modifying the surface properties of these composites to improve their functionality in specific applications, such as food packaging, where moisture resistance and antimicrobial activity are paramount [25, 26].

Due to their characteristic properties, aerogels have been employed in diverse fields such as food, chemistry, medicine, agriculture, sanitation, electronics, and construction [27]. In the food industry, they can be used in packaging as moisture absorbers or carriers of flavors, natural or synthetic additives, or nutraceuticals [4, 22, 28–31]. In agriculture, they can form delivery systems for biocides [18]. In the medical field, aerogels are used for the controlled delivery of medicines and tissue engineering [19, 32]. In chemistry, aerogels serve as carriers for catalysts and adsorbents or are used in biosensors due to their high surface area [7] In sanitary applications, they can adsorb and degrade pollutants [33, 34]. Aerogel's low weight and thermal conductivity are valuable for constructing insulation [35]. They are also used in electrical and optical equipment in electronics due to their low dielectric constant and refractive index [27, 33].

Most studies on bio-based aerogels, or bioaerogels, focus on applications in the medical field and continue to expand, driven by advancements in formulation and processing techniques. One of their primary uses is as drug carriers, as the porous network facilitates both high-loading capacities and controlled release profiles for bioactive compounds. This capability has proven valuable in pharmaceutical applications, where hydrophobic drugs can be effectively encapsulated in the aerogel matrix, enhancing their bioavailability and enabling sustained release, which is critical for therapeutic efficacy [36, 37]. Starch-based aerogels can encapsulate various therapeutic agents, including anti-cancer drugs and anti-inflammatory compounds, allowing for targeted delivery. The ability to control the release profile is particularly beneficial in treatments requiring sustained drug levels over time, reducing the need for frequent dosing. This controlled-release mechanism can lead to improved patient compliance and therapeutic outcomes. In biomedical engineering, starch-based aerogels are being investigated as scaffolds for tissue regeneration, particularly in bone and cartilage repair [23]. Their biocompatibility and porous structure support cell adhesion, proliferation, and differentiation, making them promising materials for tissue engineering applications.

Research focuses on tailoring the mechanical properties and degradation rates of these scaffolds to match the requirements of specific tissue types, promoting effective integration with surrounding tissues. Moreover, the potential to incorporate bio active molecules into the aerogel matrix allows for the formation of scaffolds that actively promote healing, adding to their appeal in regenerative medicine [12, 15].

Furthermore, starch aerogels are notable for their thermal insulating properties, a feature attributed to their highly porous structure that traps air and minimizes thermal conductivity. This characteristic position starch-based aerogels as viable materials in thermal insulation, with applications extending from construction materials to refrigerated packaging for temperature-sensitive products [2, 10]. The low thermal conductivity of starch-based aerogels makes them ideal for thermal insulation in energy-efficient buildings, where they can significantly reduce heating and cooling costs. Additionally, their lightweight nature is advantageous for reducing the overall weight of structures, contributing to energy savings during transportation and installation. Using starch-based aerogels in construction materials can lead to more sustainable building practices, aligning with global efforts to reduce the construction industry's carbon footprint.

The adaptability of starch-based aerogels has also led to their application in energy storage, particularly as templates for carbon aerogels. Carbonizing starch aerogels allows the creation of materials with high surface area and conductivity, making them suitable for supercapacitors and CO_2 adsorption applications [38, 39].

This has been instrumental in environmental applications such as air purification and renewable energy technologies, highlighting the aerogel's potential as a sustainable material for various industries. The ability of starch-derived carbon aerogels to store energy efficiently is crucial in developing next-generation energy storage systems, where quick charge and discharge cycles are needed. The porous structure of these carbon aerogels facilitates ion movement, enhancing the performance of supercapacitors, which are essential in applications ranging from electric vehicles to renewable energy systems [3, 10]. Moreover, incorporating starch-derived carbon aerogels into energy storage devices can lead to lightweight, efficient solutions that meet the increasing demand for sustainable energy technologies.

Starch-based aerogels also offer significant potential in environmental applications. Their adsorption capacity, combined with their biodegradability, positions them as suitable materials for water purification systems. Modified starch aerogels have demonstrated effective adsorption of heavy metals and organic pollutants, making them promising candidates for eco-friendly filtration technologies [36]. The ability to tailor the surface properties of starch-based aerogels enhances their efficacy in removing specific contaminants from water, making them adaptable to different environmental conditions. As environmental regulations tighten and the demand for sustainable water treatment solutions increases, starch-based aerogels present a compelling alternative to traditional materials. Research continues to optimize the adsorption characteristics of starch-based aerogels, including functionalized nanoparticles that enhance their capacity to bind specific pollutants. The potential to create composite materials that combine the strengths of starch aerogels with other adsorbents, such as activated carbon, could lead to even more effective purification systems. Additionally, starch-based aerogels are being investigated in the context of carbon capture and storage [3, 13, 14, 38]. Their high surface area and porosity make them suitable candidates for capturing carbon dioxide from industrial emissions. By modifying their surface chemistry, researchers aim to enhance their affinity for CO_2, potentially leading to the development of efficient and sustainable carbon capture technologies. In a world increasingly focused on reducing greenhouse gas emissions, innovative materials such as starch-based aerogels are crucial in carbon capture strategies [40]. This potential aligns with global efforts to mitigate climate change and transition toward a more sustainable future.

The application of aerogels in the food industry is still in its early stages but holds significant potential. In this field, starch-based aerogels serve as encapsulants for vitamins, probiotics, and flavors, offering potential in controlled-release food packaging where prolonged freshness and gradual nutrient delivery are desired [41–46]. The encapsulation of sensitive compounds in starch-based aerogels protects them from degradation during processing and storage, ensuring they retain their functionality and efficacy, improve the stability of pigments [44], and control bacteria quorum sensing [47]. This capability enhances the nutritional profile of food products and extends their shelf life. Alternatively, starch aerogels can be used to remove undesired compounds, increasing the product's shelf life or enhancing its sensory properties [45, 46]. They can also be used to increase the bioaccessibility of a nutrient molecule through impregnation into the aerogel's matrix [43, 48].

Aerogels can benefit food products by acting as moisture or exudate absorbers or controllers, preventing contact between free water and the food, thus inhibiting microbial growth, and serving as a water vapor source in dry environments [49, 50].

Despite the advantages offered by starch-based aerogels, several challenges remain, particularly in terms of scalability and material stability in high-humidity environments. Starch's inherent hydrophilicity limits its applications in humid conditions, as water absorption can lead to swelling, reduced mechanical stability, and eventual degradation. This characteristic has spurred research into various hydrophobic modifications, such as incorporating fatty acids or other hydrophobic agents, to mitigate water uptake and enhance durability [11, 51].

Hydrophobic treatments can significantly improve the moisture resistance of starch-based aerogels, enabling their use in applications where exposure to moisture is unavoidable [40]. Researchers are also exploring surface modifications, such as applying hydrophobic coatings, to enhance the aerogel's barrier properties. These modifications can involve chemical treatments or the incorporation of hydrophobic nanoparticles that improve the material's water resistance [10, 52].

Cross-linking obtained through chemical or physical methods has also been explored to improve starch-based aerogels'mechanical and thermal stability. Cross-linked starch aerogels exhibit higher resistance to moisture and maintain their structure better than non-cross-linked versions, which is crucial for applications in environments with variable humidity levels [13, 15, 36].

In addition to addressing hydrophilicity, efforts are ongoing to reduce the cost and complexity of aerogel production processes. While supercritical CO_2 drying produces aerogels with high pore integrity, it is an energy-intensive process with high operational costs and requires an additional solvent exchange step, which poses a barrier to large-scale production [38]. Researchers have investigated alternative drying methods, including vacuum and freeze-drying, as potential solutions. Freeze, vacuum, and air drying are sometimes explored as cost-effective alternatives but affect the aerogel's structure [5, 12]. Air drying of biopolymers tends to collapse the porous structures. On the other hand, vacuum [15] and freeze-drying preserve the system's porous characteristics [1]. Freeze drying can increase pore size and reduce the surface area of the processed materials due to the formation of ice crystals, compared to their supercritical dried counterparts. As a result, freeze-dried aerogels are indicated for applications as moisture absorbers and in quicker controlled delivery systems [1].

The quest for more cost-effective drying methods has led to innovations in processing techniques, such as developing hybrid drying systems that combine various methods to achieve optimal results. This approach may enhance the mechanical properties and uniformity of the resulting aerogels, making them more commercially viable [11, 16, 36]. Optimizing the gelatinization and retrogradation processes can improve material properties without significantly increasing production costs [15, 53].

As aerogel research progresses, collaboration among researchers, industry stakeholders, and policymakers will be vital to addressing challenges and unlocking the full potential of starch-based aerogels [14]. Continued investment in research and development will enable the optimization of production processes, the exploration of novel applications, and the enhancement of material properties. Furthermore, as sustainability becomes a critical consideration in material science, starch-based aerogels exemplify how natural polymers can create functional materials that meet modern societal needs. Their versatility and eco-friendly characteristics position them as key players in the future of sustainable materials, with the potential to revolutionize multiple industries, from packaging to construction and healthcare. The potential of starch-based aerogels extends beyond their immediate applications in various industries. They represent a paradigm shift towards more sustainable and environmentally friendly materials. By harnessing the unique properties of starch and integrating advanced processing techniques, researchers are paving the way for innovative solutions that align with global sustainability goals. As the body of research surrounding starch-based aerogels continues to grow, their impact on material science and sustainable development is poised to be profound and lasting.

The current chapter aims to revise the novel insights about the production and application of starch-based aerogels.

2.2 Production Methods of Starch Aerogels

Starch-based aerogels are commonly produced by two steps: hydrogel formation (gelatinization and retrogradation of starch) and drying, as shown in Fig. 2.1 [11, 22].

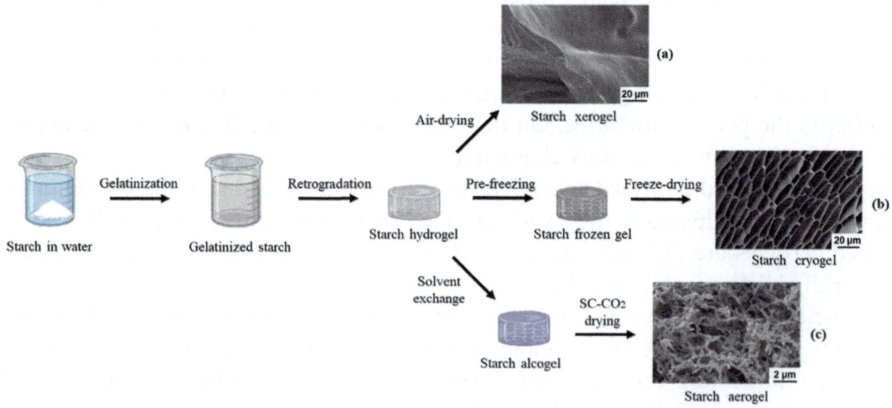

Fig. 2.1 Schematic illustration of the main steps in the production of starch-based aerogels using different solvent removal methods: (**a**) air-drying, (**b**) freeze-drying, and (**c**) supercritical CO_2 drying. (Adapted from Ubeyitogullari and Ciftci [54] and Zou et al. [55])

2.2.1 Hydrogel Formation

Starch gelatinization is a process characterized by irreversible changes, including the disruption of granular and molecular order (crystallinity loss) and the leaching of amylose molecules when starch and water mixtures are heated (up to 140 °C) [2]. Upon cooling (down to 4 °C) and storing, gelatinized starch undergoes retrogradation, in which a starch hydrogel structure, consisting of a three-dimensional polymeric network, forms due to the reassociation of amylose and amylopectin chains and partial recrystallization [2, 56]. However, fabricating starch hydrogels without the retrogradation route is also possible from the inclusion complexes between amylose and fatty acids (e.g., sodium palmitate, myristic acid) [57–59]. Hydrogels are formed as the pH of dispersions of amylose-fatty acid complexes is lowered, thereby decreasing electrostatic repulsions between the complexed anionic carboxylate head groups of sodium palmitate and allowing the amylose molecules to arrange and entangle themselves [57].

The amylose/amylopectin ratio, the concentration of starch suspension, and the gelatinization temperature are important process parameters influencing the hydrogel formation and, consequently, the physical properties of starch-based aerogels (Table 2.1).

Increasing amylose content leads to a more developed supramolecular structure and a stronger hydrogel network due to the linear molecular morphology of amylose chains that tend to reassociate easier than amylopectin after gelatinization, producing a more organized three-dimensional structure. Thus, starch aerogels prepared from a higher amylose-to-amylopectin ratio may possess a lower density due to less material shrinkage throughout processing [58, 60]. Moreover, higher amylose content results in a greater specific surface area of starch aerogels [15, 58, 60, 66]. However, pure amylose does not form starch aerogel [58].

Very low starch concentrations cannot produce stable hydrogels [70]. On the other hand, increasing starch concentration contributes to creating strong gels due to a more cross-linked network structure, resulting in less shrinkage of aerogels [54, 70]. However, very high starch concentrations may lead to high density and low porosity of aerogels or even hamper the gel handling due to the very high viscosity of starch suspension [41, 54, 65, 70].

The gelatinization temperature is another critical factor in fabricating starch-based aerogels, significantly influencing their density and specific surface area [11]. Aerogels of lower density have been obtained at higher gelatinization temperatures, which can be attributed to a higher gelatinization degree and the formation of a more fibrous starch network [12, 54, 61, 67]. Increasing the gelatinization temperature up to 120 °C increases the specific surface area of aerogels since remnants of starch granules are reduced, and a better network structure is formed. However, further increasing the gelatinization temperature results in a lower specific surface area because the swollen starch granules burst, and the structure collapses [54, 61, 67].

Table 2.1 Production of starch-based aerogels and their properties

Starch source	SCo (%)	AC (%)	Additive component	Starch hydrogel formation		Starch aerogel formation		Starch aerogel properties				References
				T_{gel} (°C)	T_{ret} (°C) and time (h)	Pre-treatment	Drying method	ρ (g cm^{-3})	SSA (m^2 g^{-1})	PS (nm)	ϕ (%)	
Corn	5.6 (w/v)	70	Sodium palmitate	140	Ni	Solvent exchange	SC-CO_2	0.120–0.185	313–362	Ni	Ni	[57]
							Air-drying	1.25	<2–3			
Corn	8 (w/w)	80	Ni	130–140	4, 48	Solvent exchange	SC-CO_2 drying	0.14	254	Ni	Ni	[58]
Corn	10 (w/w)	27	Chitosan	120	4, 168	Solvent exchange	SC-CO_2 drying	0.136–0.220	73.5–86.8	10.1–11.3	Ni	[60]
		55			4, 48			0.120–0.160	167.6–203.6	11.4–12.2		
		72						0.120–0.143	199.8–219.7	12.0–13.6		
Corn	15 (w/w)	52.6	Ni	95	4, 48	Solvent exchange	SC-CO_2 drying	1.38	66	Ni	Ni	[61]
				120				1.35	112			
				140				1.17	40			
Corn	10 (w/w)	70	Agar or microcrystalline cellulose	90	4, 24	Solvent exchange	SC-CO_2 drying	0.187–0.214	120–149	Ni	92–95	[12]
				110				0.145–0.188	169		92–93	
	15 (w/w)			90				0.251–0.278	Ni	Ni	89–92	
				110				0.201–0.230			90–92	
Corn	15 (w/v)	31.2	Ni	100	4, 48	Solvent exchange	Air-drying	Ni	21.0	Ni	Ni	[13]
							Vacuum drying		17.3			
			Sodium trimetaphosphate, adipic acid, sodium Hexametaphosphate, epichlorohydrin		Ni		Air-drying		15–6.29.6			
							Vacuum drying		20.4–22.9			

Corn	7 (w/v)	27	Trisodium citrate	90	Ni	Freezing	Freeze-drying	0.079–0.2166	183.5–403.0	2.3–2.6	72–90	[36, 62]
Corn	15 (w/v)	13.32	Calcium chloride, glycerol	93	4, 48	Solvent exchange	$SC\text{-}CO_2$ drying	0.36–0.41	Ni	Ni	75.9–69.9	[46]
Corn	14% (w/v)	27–30	Ni	110	4, 48	Solvent exchange	Vacuum drying	Ni	75.6	15.3	Ni	[63]
Corn	10 (w/w)	70	Ni	120	4, 48	Solvent exchange	$SC\text{-}CO_2$ drying	0.16	220.5	7.4	89.4	[43]
Corn	25 (w/w)	Ni	Ni	90	25, Ni	Ni	Freeze-drying	Ni	782	2	Ni	[64]
			Potassium hydroxide	Ni	Ni				807–1349	2.15–2.32		
Potato	8 (w/w)	20	Ni	130–140	4, 48	Solvent exchange	$SC\text{-}CO_2$ drying	0.23	85	Ni	Ni	[58]
Potato	5, 8, 11 (w/w)	1	Ni	95	4, 24–720	Solvent exchange	$SC\text{-}CO_2$ drying	0.21–0.35	45–141	Ni	Ni	[15]
							Vacuum drying	0.21–0.49	21–132			
						Freezing	Freeze-drying	0.16	5			
Potato	1 (w/v)	Ni	Konjac glucomannan, gelatin, wheat straw	90	4, 2	Freezing	Freeze-drying	0.020–0.039	Ni	Ni	93.8–97.2	[65]
	2 (w/v)							0.041–0.044			93.3–94.5	
	3 (w/v)							0.047–0.052			92.5–94.3	

(continued)

Table 2.1 (continued)

Starch source	SCo (%)	AC (%)	Additive component	Starch hydrogel formation		Starch aerogel formation		Starch aerogel properties				References
				T_{gel} (°C)	T_{ret} (°C) and time (h)	Pre-treatment	Drying method	ρ (g cm^{-3})	SSA (m^2 g^{-1})	PS (nm)	ϕ (%)	
Wheat	5 (w/w)	25	Ni	100	4, 48	Solvent exchange	$SC\text{-}CO_2$ drying	0.06	49.4	22.7	Ni	[54]
				120				0.05	53.2	19.2		
				130				0.04	48.7	23.5		
				140				0.03	45.4	21.5		
	10 (w/w)			100				0.12	48.4	21.2		
				120				0.12	54.1	20.3		
				130				0.11	35.2	25.9		
				140				0.10	40.5	25.1		
	15 (w/w)			100				0.37	20.2	24.8		
				120				0.29	58.3	21.8		
				130				0.18	44.1	22.4		
				140				0.15	40.5	20.9		
Wheat	10 (w/w)	25	Ni	120	4, 48	Solvent exchange	$SC\text{-}CO_2$ drying	0.11	61.5	19	92.5	[66]
Wheat	10 (w/w)	25	Ni	120	4 °C, 48 h	Solvent exchange	$SC\text{-}CO_2$ drying	0.11	60.9	17.6	92.5	[67]
				130				0.1	34.7	25.1	92.9	
				120			Air-drying	0.41	0.04	267.9	72.6	
Waxy maize	5 (w/w)	1	Ni	95	4, 72–264	Freezing	Freeze-drying	0.07	3–5	Ni	95	[55]
	8 (w/w)							0.11			93	
	11 (w/w)							0.15			90	

Waxy maize	5, 8, 11 (w/w)	1	Ni	95	4, 24–720	Solvent exchange	$SC\text{-}CO_2$ drying	0.44	17	Ni	Ni	[15]
							Vacuum drying	0.31–0.63	6–87			
						Freezing	Freeze-drying	0.15	5			
Waxy potato	8 (w/w)	20	Ni	130–140	4, 163	Solvent exchange	$SC\text{-}CO_2$ drying	0.20	88	Ni	Ni	[58]
Acetylated potato	1 (w/v)	Ni	Cellulose	45	Ni	Ni	Freeze-drying	18.42	Ni	95.6	87	[41]
	3 (w/v)							21.86		52	85	
	5 (w/v)							32.07		37.5	78	
	7 (w/v)							46.63		29.9	69	
	9 (w/v)							54.77		24.7	64	
Pea	8 (w/w)	40	Ni	130–140	4, 48	Solvent exchange	$SC\text{-}CO_2$ drying	0.14	221	Ni	Ni	[58]
Pea	5, 8, 11 (w/w)	33–36	Ni	95	4, 24–720 days	Solvent exchange	$SC\text{-}CO_2$ drying	0.10–0.11	166–253	Ni	Ni	[15]
							Vacuum drying	0.51–1.21	64–150			
Pea	5 (w/v)	Ni	Ni	95	25, 2	Freezing	Freeze-drying	0.032–0.099	4.88–17.48	2.49–2.99	97–99	[14]
Cassava	6.7 (w/v)	Ni	Polyvinyl alcohol	70	Ni	Freezing	Freeze-drying	0.13–0.15	Ni	Ni	Ni	[68]
Lentil	5 (w/v)	Ni	Ni	95	25, 2	Freezing	Freeze-drying	0.079–0.110	5.31–7.47	2.25–3.70	97–99	[14]

(continued)

Table 2.1 (continued)

Starch source	SCo (%)	AC (%)	Additive component	Starch hydrogel formation		Starch aerogel formation		Starch aerogel properties				References
				T_{gel} (°C)	T_{ret} (°C) and time (h)	Pre-treatment	Drying method	ρ (g cm^{-3})	SSA (m^2 g^{-1})	PS (nm)	ϕ (%)	
Faba bean	5 (w/v)	Ni	Ni	95	25, 2	Freezing	Freeze-drying	0.07–0.09	2.26–3.81	2.38–4.68	97–99	[14]
Defatted rice bran	15 (w/w)	Ni	Ni	95	4, 48	Solvent exchange	SC-CO_2 drying	0.27–0.29	36–47	16–19	83–85	[69]
Yam	10 (w/v)	Ni	Inactivated yeast	90	4, 48	Freezing	Freeze-drying	Ni	4.12	20.48	Ni	[42]

SCo starch concentration, *AC* amylose content, T_{gel} gelatinization temperature, T_{ret} retrogradation temperature, ρ density, *SSA* specific surface area, *PS* pore size, ϕ porosity, *Ni* not informed

Adding gelling improvers into starch suspensions is another potential way to improve the mechanical properties and introduce new functionalities of starch-based aerogels [11]. Cross-linking agents, such as sodium metaphosphate [13, 67], adipic acid [13], epichlorohydrin [13], trisodium citrate [62, 71], and calcium chloride [46], can be used to modify the microstructure of starch-based aerogels. The hydroxyl groups on starch molecules can react to cross-linking agents, forming an interconnected gel network [11, 13]. Cross-linking of starch increases the uniformity of pore distribution of aerogels [2, 13, 67, 71]. Furthermore, the employment of non-starch biopolymers (e.g., agar, cellulose, chitosan, gelatin) in the gelling system can also significantly affect the morphology and properties of starch aerogels (Table 2.1) [2, 11].

2.2.2 *Drying Procedure*

The resulting starch hydrogels are then dried to form starch aerogels (Fig. 2.1). This step involves replacing the liquid in the hydrogel structures with gas. Drying techniques must be carefully selected to maintain the integrity of the original network structure and avoid the shrinkage of starch-based aerogels. The most widely used to produce starch aerogels are air-drying, freeze-drying, and supercritical drying [11, 22].

Air-drying starch hydrogels (to produce xerogels) is a liquid evaporation process performed at room temperature and ambient pressure until a constant weight is achieved. Although it is a simple and economical method, air-drying has caused the collapse of gel pores, structural shrinkage, and formation of cracks and macropores (Table 2.1), or even non pores (Fig. 2.1a) due to the capillary forces from the high surface tension of solvent in small gel pore sizes during drying [13, 57, 63, 67]. Air-dried starch aerogels have shown a higher density and lower specific surface area and porosity than other dried starch aerogels [57, 67]. Reducing drying pressure (vacuum drying) can be an alternative technique for improving the properties of starch xerogel, such as density, specific area, and morphology [15, 63]. During vacuum drying, the surface tension is reduced, thus preventing pore collapse and structural shrinkage [63].

Freeze-dying of starch gels (to produce cryogel) removes the solvent from the frozen gel by sublimating ice crystals under vacuum. There is no capillary pressure existing in the pores during this drying process. However, ice crystal formation during pre-freezing may damage the resulting aerogel structure, decreasing specific surface area and forming macropores [15, 53, 57, 63]. The growth of ice crystals during freezing pushes aside the pore walls of starch gel [53]. In addition, freeze-drying is a costly method because of its extended process time and high energy consumption.

Supercritical CO_2 (SC-CO_2) drying has been the most widely applied technique for producing starch-based aerogels (Table 2.1). The use of supercritical CO_2 as a supercritical drying fluid is related to its properties, such as non-toxicity and chemical inertness, and its mild operating conditions to reach the critical point (7.4 MPa, 31.1 °C) [11, 72]. Since the water in hydrogel pores has a low affinity for CO_2, a solvent exchange is required to replace the water with a CO_2-miscible solvent (usually ethanol) before SC-CO_2 drying [22, 57]. Solvent exchange is commonly performed by soaking the starch hydrogel for a determined time in a consecutive series of water-new-solvent solutions with increasing new solvent concentrations [72]. As capillary pressure is absent during SC-CO_2 drying, the porous structure of starch hydrogel is preserved [2]. Starch aerogels produced by SC-CO_2 drying have presented a high specific surface area and porosity [15, 54]. Nevertheless, the application of this drying technique for large-scale industrial production of aerogels is still restricted due to its high energy consumption and solvent exchange steps [22].

2.2.3 Fabrication Routes

Starch-based aerogels have been prepared in different forms, such as monoliths, beads, and microspheres, with sizes ranging from nanoscale to millimeters [11]. Monoliths (Fig. 2.2a) are the primary manufacturing options of starch-based aerogels [12, 54, 55, 65, 69], in which gelatinized starch suspensions are poured into molds (commonly cylindrical), followed by gelation and drying [11]. The dripping

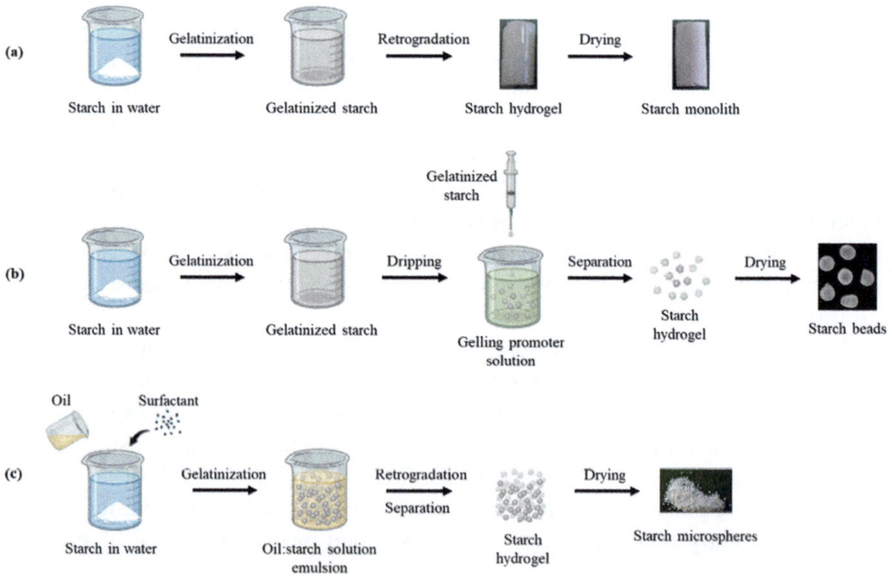

Fig. 2.2 Schematic illustration of the processing routes of starch-based aerogels: (**a**) monoliths, (**b**) beads, and (**c**) microspheres. (Adapted from Kenar et al. [57], García-González et al. [61], and Dogenski et al. [12])

approach usually produces starch-based aerogel beads (Fig. 2.2b). A starch solution is delivered dropwise to a gelling promoter solution (e.g., pH and/or temperature-controlled solution) using a syringe or atomizing nozzle. Starch gelation instantly occurs, and subsequently, the gel beads are dried [57, 72]. Aerogel microspheres (Fig. 2.2c) are obtained by the emulsion-gelation method. A water-in-oil emulsion is prepared by mixing starch dispersion in water and oil with a surfactant. Upon heating of the resulting emulsion, gel spherical particles are formed. After gelation, microspheres are separated (through centrifugation or filtration) from the oil phase and then dried [11, 61].

2.3 Food Applications of Starch Aerogels

Starch aerogels have been applied to some foodstuffs (Table 2.2), such as granola bars [43], tea [42], cheese [41], pork meat [45], and spinach [46]. In general, these aerogels are used to either release an active compound or to remove a non-desired compound, therefore increasing the product's shelf life [45, 46]. It is also possible to increase the bioaccessibility of a molecule through impregnation into the aerogel's matrix [43, 48]. Interestingly, the addition of aerogels in a product may have no or minimal impact on textural properties [42, 43].

Regarding the release of active compounds, *Thymus daenensis* Celak essential oil and copper nanoparticles were impregnated into starch aerogel and applied in cheese packaging and pork meat, respectively [41, 45]. In the case of cheese, the essential oil was slowly released in the package's atmosphere during storage at 7 °C. The concentration of psychrophiles was reduced, while those of yeasts and molds was maintained. Moreover, it was possible to eliminate *E. coli* (4.2 log CFU/g) after 7 days of storage. The efficacy was dependent on the oil's concentration in the aerogel, meaning that higher concentrations were mor effective [41].

Concerning nanoparticles with aerogels, applying aerogel pads containing copper nanoparticles in direct contact with pork meat extended its shelf life. The meat had pH lower than 6 and a total viable count lower than log 6 CFU/g after 14 days of storage at 4 °C. This indicates that it was still considered sub-fresh and, therefore proper for consumption. For comparison, samples without aerogel pads with copper nanoparticles were considered rotten after the same storage time, and their pH was much closer to that of rotten meat after 7 days [45].

The applications concerning the removal of undesired compounds was performed using starch-glycerol aerogel or modified starch aerogel loaded with inactivated yeast. The former was applied to control the relative humidity in packages containing spinach leaves, which extended its shelf life by 3 days and was possible due to the affinity of starch and glycerol with water [46]. The latter was used to remove zearalenone—a *Fusarium* mycotoxin—from corn silk tea, performing better than the removal using non-embedded inactivated yeast. Curiously, aerogels made only with starch modified by pullulanase, although less effective, also presented the potential to adsorb zearalenone from cornsilk tea [46].

Table 2.2 Applications of starch aerogels in foods

Food application	Ingredients	Production method	Major findings	References
Controlled release of essential oil in cheese packaging	Cellulose Acetylated potato starch *Thymus daenensis* Celak essential oil	Freeze-drying	Aerogels were prepared by immersion in solutions containing 25 and 50 times the minimum inhibitory dose of *Thymus daenensis* Celak essential oil. Aerogels were attached to the lid of cheese packaging. Those made with higher concentration of essential oil were able to reduce the concentration of psychrophiles and better control the growth of yeast and mold. A lower dosage was only able to maintain the level of psychrophiles and received higher scores in sensory analysis. *E. coli* population in cheese was reduced from 4.2 to 0 log CFU/g in 7 and 14 days with aerogels with lower and higher essential oil content, respectively	[41]
Removal of zearalenone from corn silk tea	Modified yam starch STMP Citric acid α-amylase Pullulanase	Freeze-drying	Loaded SZ4 aerogel was able to adsorb zearalenone from corn silk tea without changes to clarity and reducing sugar content. However, similarly to free SZ4, tea's pH and total phenols was reduced after adsorption. Also, cell viability of 293 T, TM3 and L02 was higher when adsorption of zearalenone was performed using the loaded aerogel	[42]
Improve bioaccessibility of phytosterols in pudding and granola bars	Wheat starch Phytosterol nanoparticles	SC-CO_2 drying	The use of aerogels did not affect the hardness of granola bars with low or regular-fat formulations. Moreover, puddings made with crude phytosterol, and starch aerogels had lower hardness and gumminess. These textural changes were due to differences in physical properties between starch aerogel and pregelatinized starch. The bioaccessibility of phytosterols was drastically increased when first loaded into starch aerogel in all granola bar formulations and in puddings. Dissolution of phytosterol was increased in both gastric (from 1% to 14%) and intestinal fluids (from 1% to 6%) after impregnation in starch aerogel. Impregnation also increased the solubility of phytosterol in water (from 0.0003 to 0.0112 mg/mL). This higher dissolution is linked to the formation of nanosized amorphous particles during SC-CO_2 impregnation	[43, 48]

Controlled release of copper nanoparticles during shelf life applied in pork meat mats	Chitosan Dialdehyde starch Soy lecithin Cholesterol Copper nanoparticles	Freeze-drying	Aerogel pads were able to keep pork meat's pH in sub-fresh range even after 14 days of storage at 4 °C. However, the total viable count was higher than 6 Log CFU/g, indicating it was not safe for consumption. On the other hand, the addition of copper nanoparticles to the aerogel structure enhanced the antimicrobial effect. Pork meat had only 5 Log CFU/g after 14 days of storage at 4 °C, implying it was still safe	[45]
Moisture absorber for spinach leaves	Corn starch $CaCl_2$ Glycerol	SC-CO_2 drying	Using glycerol containing-starch aerogel increased the shelf life of spinach leaves by 3 days both in room and refrigerating temperature. Glycerol improved the water absorption capacity of corn starch aerogels by enhancing the number of sites where water can bond	[46]

The nutritional value of foods can be improved by loading a starch aerogel with a target molecule. For example, granola bars can be fabricated using phytosterol-loaded starch aerogels without impacting textural properties. This addition improves the capacity of a product to regulate cholesterol, as the consumption of phytosterols is linked to a reduction in LDL-cholesterol levels [43]. The increase in bioavailability was linked to the formation of amorphous phytosterol structures inside the starch aerogel matrix during impregnation. This amorphous morphology is easier to digest than the crystalline structure of free phytosterols [48].

Although they have not applied aerogels to foodstuff, some studies show promising applications (Table 2.3). The potential applications are similar to those aforementioned, with most applying aerogels to either make a controlled release of active compounds or to absorb water [49, 50]. Yet, it is possible to use aerogels to improve the stability of pigments [44] and control bacteria quorum sensing [47].

Pigments may have their photostability improved when adsorbed into starch aerogels. A solution of β-carotene degrades after 4 h of exposure to UV-light, whereas the impregnation in starch aerogel resulted in 27% of the initial concentration of pigment remaining after 48 h of exposure. Similar improvements can be seen for when exposed to light [44]. This increased stability is most likely due to the physical protection to light promoted by the starch chains of the aerogel. Other light sensitive pigments that were embedded in starch aerogels may also benefit from improved photo stability, such as curcumin [47, 73] and anthocyanins [50]. Therefore, starch aerogels could be a way to increase the use of natural pigments in food formulations and substitute the commonly used synthetic pigments.

Apart from possible increased stability, curcumin can also be used alongside starch aerogels for modulating quorum sensing of bacteria [47]. The use of this combination resulted in a decrease of the production of numerous pigments linked to the metabolism of gram-negative bacteria, such as *P. aeruginosa* and *C. violaceum*. Moreover, it can also inhibit the formation of biofilm and disrupt those that are already formed [47].

Impregnation of copper nanoparticles [45] and essential oil [41] in aerogels can indeed be beneficial for foodstuff and it can prolong its shelf life by slowly releasing these active compounds. In this way, there are other possible molecules that can be embedded into starch aerogels and may contribute to food quality. Impregnation of molecules from pinhão coat [49] and from grape skin [50] may contribute to the antioxidant content of foods and prevent the oxidation of other compounds, such as vitamins and lipids. Also, adsorption of trans-2-hexenal [71] and quercetin [74] into starch aerogels resulted in a slower release of these molecules, which can prolong their activity.

These studies show that starch aerogels can indeed be effectively applied to foods and may enhance their properties, such as shelf life and nutritional value. However, few studies have applied them in foods, with applications being focused on the use of aerogels to manufacture active packages. Hence, more studies should apply aerogels in foods, as they clearly display potential to enhance their properties. This way, the effectiveness of starch aerogels will be clearer.

Table 2.3 Possible applications of starch aerogel in foods

Possible application	Ingredients	Production method	Major findings	References
Improved stability of carotenoids	Corn starch Sodium hexametaphospate β-carotene	Freeze-drying	Encapsulation of β-carotene in starch aerogel improved its photostability at room temperature when compared to its raw solution. When exposed to UV-light, 27% of pigment remained in the aerogel after 48 h, while the pigment's solution had almost completely degraded after 4 h of exposure. Regarding exposure to light, 2 and 38% of β-carotene remained after 4 weeks of exposure for the pigment in solution and encapsulated, respectively. β-carotene was rapidly released in the first 60 min of *in-vitro* digestion	[44]
Modulate quorum sensing of Gram-negative bacteria	Starch Curcumin	$SC\text{-}CO_2$ drying	Minimum inhibitory concentration (MIC) was 8 mg of curcumin-loaded aerogel/mL. Concentrations lower than MIC were able to reduce the production of violacetin, prodigiosin, and pyoverdin, and reduced biofilms	[47]
Controlled release of active compounds from pinhão (*Araucaria angustifolia*) coat	Native corn starch Anionic corn starch Pinhão coat extract	Freeze-drying	The antioxidant capacity of aerogels made with anionic starch was higher than native starch. A higher concentration of pinhão coat extract led to greater antioxidant capacity. No difference in the release rate of active compounds was observed between anionic and native starch aerogels. The better active ability of anionic aerogel may be linked to a synergistic effect between the extract and phosphate groups of the matrix, where phosphate groups enhance the sequestering ability of polysaccharides	[49]
Moisture absorber and controlled release of active compounds from grape skin	Germinated wheat starch Polyethylene oxide Grape skin extract	Freeze-drying	Higher concentrations of grape skin extract led to higher inhibition of DPPH and ABTS radicals. The addition of polyethylene oxide in the aerogels improved the antioxidant capacity, and water absorption, and promoted a slower release of antioxidant molecules in both hydrophilic and hydrophobic medium	[50]

(continued)

Table 2.3 (continued)

Possible application	Ingredients	Production method	Major findings	References
Controlled release of curcumin in gastric and intestinal conditions	Corn starch Chitosan Acetic acid Curcumin	SC-CO_2 drying	Loading curcumin into starch aerogel via SC-CO_2 improved its release in gastric conditions when compared to mixing curcumin powder with aerogel powder. Also, the addition of chitosan in the aerogel formulation further increased curcumin release in gastric conditions. Similar behavior was observed in intestinal conditions. This increased release was linked to the surface positive charge of chitosan, which can disrupt hydrogen bonds between curcumin and starch, and to the effect of pancreatic amylotyc enzymes during intestinal fase, causing breakage of starch chains. It is estimated that loading curcumin into starch-chitosan aerogels improved inaccessibility by 20–30 times	[73]
Controlled release of trans-2-hexenal	Starch Trisodium citrate Trans-2-hexenal Span 80	Freeze-drying	Cross-linking with trisodium citrate resulted in an aerogel with more uniform pores. The release of trans-2-hexenal from this cross-linked aerogel was much slower than that of free trans-2-hexenal. Additionally, aerogel coating with sorbitan monooleate (Span 80) resulted in an even slower release of trans-2-hexenal. Coating resulted in slower inactivation of *A. parasiticus* spores when compared to non-coated aerogel. The slower release is attributed to the intrastructural barriers and the closure of openings by the surfactant	[71]
Controlled release of quercetin	Maize starch Quercetin	SC-CO_2 drying	Diffusion of quercetin into a food simulant was reduced when the compound was incorporated into aerogels. Moreover, quercetin release was slower when absorbed into maize starch aerogel compared to calcium alginate aerogel. This slower dissolution is attributed to the lower diffusion of quercetin in maize starch aerogel	[74]

2.4 Conclusions

Aerogels stand out as advanced materials due to their unique combination of high surface area, low density, and high porosity, making them valuable in diverse applications such as construction, electronics, and food packaging. Integrating these characteristics with the inherent advantages of starch—a bio-based, biodegradable, biocompatible, non-toxic, and cost-effective polymer—unlocks new opportunities for advancing aerogel technologies.

The adaptability of starch modification processes further enhances its utility in aerogels. These modifications may enable the fine-tuning of properties such as water resistance, thermal stability, and controlled release or adsorption, broadening the scope of potential applications. In this context, combining starch with other polymers or materials, such as cellulose or composites, represents a promising research avenue for developing multifunctional aerogels tailored to specific needs.

However, the production of high-quality aerogels often involves energy- and time-intensive processes like supercritical carbon dioxide (SC-CO_2) drying and freeze-drying, which contribute to their high cost. As a result, their current applications may be more economically viable in high-value fields such as medicine and pharmaceuticals. To expand the use of starch-based aerogels into lower-cost products, such as disposable food packaging, research focusing on process optimization and the development of more cost-effective manufacturing techniques is essential. Achieving cost reductions could benefit the aerogel technology by enabling its adoption across a broader range of industries.

One particularly compelling application is in disposable thermal insulating food packaging. Starch aerogels show potential as eco-friendly alternatives to conventional polymer packaging, such as expanded, especially for cold and fat-containing products like ice cream. However, their sensitivity to temperature and moisture necessitates further exploration of strategies to enhance water resistance and thermal stability. Innovations in this area could lead to practical and sustainable solutions for diverse packaging requirements.

In food and food packaging, starch aerogels have already demonstrated significant promise, contributing to food preservation, nutrient bioavailability, and the adsorption of undesired compounds. They function as moisture absorbers, encapsulants for stabilizing and controlling the release of bioactive or sensory compounds, and as agents for removing harmful substances. Notably, their capacity to stabilize pigments such as curcumin underscores their potential in improving food quality, though further studies in real food systems are essential to validate these effects.

Starch-based aerogels embody a versatile and sustainable platform with the potential to solve diverse demands across industries. Continued research into starch modification, composite formulations, multi-layered packaging, and cost-effective production processes will be pivotal in realizing their full potential, paving the way for innovative, environmentally conscious applications.

Acknowledgments The authors would like to thank the Federal University of Santa Catarina. R.R. Dalsasso and R.S. Simão thank the National Council for Scientific and Technological Development (CNPq) for the financial support (process numbers 175329/2023-0 and 150964/2024-2, respectively). W.D.C. Chacon gratefully acknowledges the Fundação de Amparo à Pesquisa e Inovação do Estado de Santa Catarina (FAPESC) for the Ph.D. fellowship (process number 3003/2021). G.C. Leandro acknowledges the Coordination of Superior Level Staff Improvement (CAPES) for the Ph.D. fellowship (process number 88887.963657/2024-00).

Conflicts of Interest The authors declare no conflict of interest.

References

1. Basak S, Singhal RS. The potential of supercritical drying as a "green" method for the production of food-grade bioaerogels: a comprehensive critical review. Food Hydrocoll [Internet]. 2023;141(January):108738. Available from: https://doi.org/10.1016/j.foodhyd.2023.108738.
2. Zhu F. Starch based aerogels: production, properties and applications. Trends Food Sci Technol. Elsevier Ltd. 2019;89:1–10.
3. Zhai Z, Zheng Y, Du T, Tian Z, Ren B, Xu Y, et al. Green and sustainable carbon aerogels from starch for supercapacitors and oil-water separation. Ceram Int. 2021;47(15):22080–7.
4. Manzocco L, Plazzotta S, Powell J, de Vries A, Rousseau D, Calligaris S. Structural characterisation and sorption capability of whey protein aerogels obtained by freeze-drying or supercritical drying. Food Hydrocoll. 2021;122(June 2021):107117.
5. Pantić M, Nowak M, Lavrič G, Knez Ž, Novak Z, Zizovic I. Enhancing the properties and morphology of starch aerogels with nanocellulose. Food Hydrocoll. 2024;156(June):110345.
6. Yoda S, Takeshita S, Ono T, Tada R, Ota H. Development of a new silica aerogel-polypropylene foam composite as a highly flexible thermal insulation material. Front Mater. 2021;8(May):1–9.
7. Jung DB, Song Y, Lee YR, Cha MJ, Jeong K, Choi J, et al. Quaternarized chitosan nanofiber and ZIF aerogel composites for synergetic CO_2 cycloaddition catalysis. Carbohydr Polym [Internet]. 2025;347(June 2024):122685. Available from: https://doi.org/10.1016/j.carbpol.2024.122685.
8. Guajardo S, Figueroa T, Borges J, Meléndrez M, Fernández K. Comparative study of graphene oxide-gelatin aerogel synthesis: chemical characterization, morphologies and functional properties. J Inorg Organomet Polym Mater [Internet]. 2021;31(4):1517–26. Available from: https://doi.org/10.1007/s10904-020-01770-9
9. Jiang J, Zhang Q, Zhan X, Chen F. A multifunctional gelatin-based aerogel with superior pollutants adsorption, oil/water separation and photocatalytic properties. Chem Eng J [Internet]. 2019;358(October 2018):1539–51. Available from: https://doi.org/10.1016/j.cej.2018.10.144.
10. Chen Y, Dai G, Gao Q. Starch nanoparticles-graphene aerogels with high supercapacitor performance and efficient adsorption. ACS Sustain Chem Eng. 2019;7(16):14064–73.
11. Zheng Q, Tian Y, Ye F, Zhou Y, Zhao G. Fabrication and application of starch-based aerogel: technical strategies. Trends Food Sci Technol. Elsevier Ltd. 2020;99:608–20.
12. Dogenski M, Navarro-Díaz HJ, de Oliveira JV, Ferreira SRS. Properties of starch-based aerogels incorporated with agar or microcrystalline cellulose. Food Hydrocoll. 2020;108:106033.
13. Wang Y, He M, Wu Y, Liu Y, Ouyang J. Effect of crosslinking agents on the physicochemical and digestive properties of corn starch aerogel. Starch/Staerke. 2021;73(3–4):2000161.
14. Falua KJ, Babaei-Ghazvini A, Acharya B. Comparative study of the structure and mechanical properties of starch aerogels fabricated from air-classified and isolated pulse starches. Int J Biol Macromol. 2024;257:128478.
15. Zou F, Budtova T. Starch alcogels, aerogels, and aerogel-like xerogels: adsorption and release of theophylline. ACS Sustain Chem Eng. 2023;11(14):5617–25.

16. Mottola S, Iannone G, Giordano M, González-Garcinuño Á, Jiménez A, Tabernero A, et al. Supercritical impregnation of starch aerogels with quercetin: fungistatic effect and release modelling with a compartmental model. Int J Biol Macromol. 2023;253:127406.
17. Yuan Y, Solin N. Protein-based luminescent aerogels with elastic properties. Green Chem Lett Rev [Internet]. 2022;15(3):508–18. Available from: https://doi.org/10.1080/17518253.2022.2102941.
18. Montaser AS, Abdelhameed RM, Shaheen TI. Formulating of the sustained release of tebuconazole pesticide using chitosan aerogel reinforced NFC/$CaCO_3$ nanocomposite. Int J Biol Macromol [Internet]. 2024;256(P1):128419. Available from: https://doi.org/10.1016/j.ijbiomac.2023.128419.
19. Sang Y, Gao J, Han X, Liang T, Chen T, Zhao Y. Preparation and sustained release of diatomite incorporated and Eudragit L100 coated hydroxypropyl cellulose/chitosan aerogel microspheres. Int J Biol Macromol [Internet]. 2024;267(P1):131447. Available from: https://doi.org/10.1016/j.ijbiomac.2024.131447.
20. Dhua S, Mishra P. Development of highly reusable, mechanically stable corn starch-based aerogel using glycerol for potential application in the storage of fresh spinach leaves. Int J Biol Macromol [Internet]. 2023;242(P4):125102. Available from: https://doi.org/10.1016/j.ijbiomac.2023.125102.
21. Alavi F, Ciftci ON. Green and single-step simultaneous composite starch aerogel formation-high bioavailability curcumin particle formation. Int J Biol Macromol [Internet]. 2024;264(P1):129945. Available from: https://doi.org/10.1016/j.ijbiomac.2024.129945.
22. Zhao D, Zhang X, Zhang Y, Xu E, Yan S, Xu H, et al. Recent advances in the fabrication, characterization and application of starch-based materials for active food packaging: hydrogels and aerogels. Sustain Food Technol. 2024;2(3):615–34.
23. Santos-Rosales V, Alvarez-Rivera G, Hillgärtner M, Cifuentes A, Itskov M, García-González CA, et al. Stability studies of starch aerogel formulations for biomedical applications. Biomacromolecules. 2020;21(12):5336–44.
24. Jayarathna S, Andersson M, Andersson R. Recent advances in starch-based blends and composites for bioplastics applications. PolymersMDPI. 2022;14:4557.
25. Liu D, Zhao P, Chen J, Yan Y, Wu Z. Recent advances and applications in starch for intelligent active food packaging: a review. Foods. 2022;11:2879. MDPI
26. Gamage A, Thiviya P, Mani S, Ponnusamy PG, Manamperi A, Evon P, et al. Environmental properties and applications of biodegradable starch-based nanocomposites. Polymers. 2022;14:4578. MDPI.
27. Rafieian F, Dufresne A, Askari G, Rezaei A, Seyedhosseini-Ghaheh H, Jafari SM. Aerogels as novel ingredients: production, properties and applications in medical, food and environmental sectors. Colloids Surf A Physicochem Eng Asp [Internet]. 2024;687(February):133410. Available from: https://doi.org/10.1016/j.colsurfa.2024.133410.
28. Mirmoeini SS, Hosseini SH, Lotfi Javid A, Esmaeili Koutamehr M, Sharafi H, Molaei R, et al. Essential oil-loaded starch/cellulose aerogel: preparation, characterization and application in cheese packaging. Int J Biol Macromol [Internet]. 2023;244(February):125356. Available from: https://doi.org/10.1016/j.ijbiomac.2023.125356.
29. Siddiqui SA, Singh P, Utama DT, Samatra MY, Ahmad A, Wani SA. Encapsulation of bioactive compounds in foods for diabetics - sources, encapsulation technologies, market trends and future perspectives – a systematic review. Food Bioprod Process [Internet]. 2024;147(June):277–303. Available from: https://doi.org/10.1016/j.fbp.2024.07.007.
30. Zhang P, Wang Y, Liu Y, Wu Y, Ouyang J. Improved stability of β-carotene by encapsulation in SHMP-corn starch aerogels. Food Chem [Internet]. 2023;406(June 2022):135040. Available from: https://doi.org/10.1016/j.foodchem.2022.135040.
31. Wang F, Xu Z, Chen L, Qiao Z, Hu Y, Fan X, et al. Super absorbent resilience antibacterial aerogel with curcumin for fresh pork preservation. Food Control [Internet]. 2024;159(22):110289. Available from: https://doi.org/10.1016/j.foodcont.2024.110289.

32. Iglesias-Mejuto A, Magariños B, Ferreira-Gonçalves T, Starbird-Pérez R, Álvarez-Lorenzo C, Reis CP, et al. Vancomycin-loaded methylcellulose aerogel scaffolds for advanced bone tissue engineering. Carbohydr Polym. 2024;324(October 2023);121536:1–14.
33. Zhai Z, Li H, Zheng Y, Ji Y, Peng H, Gao Y, et al. High specific surface area carbon aerogel derived from starch for methylene blue adsorption and supercapacitors. Int J Biol Macromol [Internet]. 2024;274(P1):133282. Available from: https://doi.org/10.1016/j.ijbiomac.2024.133282.
34. Wang R, Dong H, Li X, Zhou L, Zhu W, Chen T. Biomass encapsulated ZIF-8-derived ZnO carbon aerogels for efficient uranium extraction by synergistic adsorption-photoreduction. Chem Eng J [Internet]. 2023;478(October):147331. Available from: https://doi.org/10.1016/j.cej.2023.147331.
35. Abdallah HA, Tannous JH, Abu-Jdayil B. Date palm wood-derived cellulose aerogel dissolved in ionic liquids as a green thermal insulation construction material. Constr Build Mater [Internet]. 2024;436(June):136957. Available from: https://doi.org/10.1016/j.conbuildmat.2024.136957.
36. Camani PH, Midhun Dominic CD, Parra DF, Maltez HF, Rosa DS. Divalent metal ion removal from simulated water using sustainable starch aerogels: effect of crosslinking agent concentration and sorption conditions. Int J Biol Macromol. 2023;226:628–45.
37. Baudron V, Taboada M, Gurikov P, Smirnova I, Whitehouse S. Production of starch aerogel in form of monoliths and microparticles. Colloid Polym Sci. 2020;298(4–5):477–94.
38. Dias ALB, Hatami T, Viganó J, Santos De Araújo EJ, Mei LHI, Rezende CA, et al. Role of supercritical CO2impregnation variables on β-carotene loading into corn starch aerogel particles. J CO2 Util. 2022;63:102125.
39. Villegas M, Oliveira AL, Bazito RC, Vidinha P. Development of an integrated one-pot process for the production and impregnation of starch aerogels in supercritical carbon dioxide. J Supercrit Fluids. 2019;154:104592.
40. Wu K, Wu H, Wang R, Yan X, Sun W, Liu Y, et al. The use of cellulose fiber from office waste paper to improve the thermal insulation-related property of konjac glucomannan/starch aerogel. Ind Crop Prod. 2022;177:114424.
41. Mirmoeini SS, Hosseini SH, Lotfi Javid A, Esmaeili Koutamehr M, Sharafi H, Molaei R, et al. Essential oil-loaded starch/cellulose aerogel: preparation, characterization and application in cheese packaging. Int J Biol Macromol. 2023;244(February):125356.
42. Fang J, Lu J, Zhao X, Zhang T, Ye H, Fang T, et al. Zearalenone removal using inactivated yeast embedded in porous modified yam starch aerogels and its application in corn silk tea. Food Chem. 2024;460(March):140593.
43. Ubeyitogullari A, Ciftci ON. In vitro bioaccessibility of novel low-crystallinity phytosterol nanoparticles in non-fat and regular-fat foods. Food Res Int. 2019;123(November 2018):27–35.
44. Zhang P, Wang Y, Liu Y, Wu Y, Ouyang J. Improved stability of β-carotene by encapsulation in SHMP-corn starch aerogels. Food Chem. 2023;406:135040.
45. Chen L, Niu X, Fan X, Liu Y, Yang J, Xu X, et al. Highly absorbent antibacterial chitosan-based aerogels for shelf-life extension of fresh pork. Food Control. 2022;136(October 2021):108644.
46. Dhua S, Mishra P. Development of highly reusable, mechanically stable corn starch-based aerogel using glycerol for potential application in the storage of fresh spinach leaves. Int J Biol Macromol. 2023;242(P4):125102.
47. Mabood Husain F, Arshad M, Ahmad Khan R, Imran A, Ali S. Curcumin loaded starch-based aerogels interfere with quorum sensing regulated virulence functions and biofilm of bacterial pathogens. J King Saud Univ Sci. 2024;36(9):103406.
48. Ubeyitogullari A, Ciftci ON. Phytosterol nanoparticles with reduced crystallinity generated using nanoporous starch aerogels. RSC Adv. 2016;6(110):108319–27.
49. Fonseca LM, da Silva FT, Bruni GP, Borges CD, Zavareze E d R, Dias ARG. Aerogels based on corn starch as carriers for pinhão coat extract (Araucaria angustifolia) rich in phenolic compounds for active packaging. Int J Biol Macromol. 2021;169:362–70.
50. Silva FT, Fonseca LM, Bruni GP, Crizel RL, Oliveira EG, Zavareze E d R, et al. Absorbent bioactive aerogels based on germinated wheat starch and grape skin extract. Int J Biol Macromol. 2023;249(July):126108.

51. Huang J, Gao J, Qi L, Gao Q, Fan L. Preparation and properties of starch–cellulose composite aerogel. Polymers (Basel). 2023;15(21):4294.
52. Zhao YW, Tian MZ, Huang P. Starch/clay aerogel reinforced by cellulose nanofibrils for thermal insulation. Cellulose. 2021;28(6):3505–13.
53. Zou F, Budtova T. Tailoring the morphology and properties of starch aerogels and cryogels via starch source and process parameter. Carbohydr Polym. 2021;255:117344.
54. Ubeyitogullari A, Ciftci ON. Formation of nanoporous aerogels from wheat starch. Carbohydr Polym. 2016;147:125–32.
55. Zou F, Bouvard JL, Pradille C, Budtova T. Ice-templated additive-free porous starches with tuned morphology and properties. Eur Polym J. 2022;176:111403.
56. Wang S, Li C, Copeland L, Niu Q, Wang S. Starch retrogradation: a comprehensive review. Compr Rev Food Sci Food Saf. 2015;14(5):568–85.
57. Kenar JA, Eller FJ, Felker FC, Jackson MA, Fanta GF. Starch aerogel beads obtained from inclusion complexes prepared from high amylose starch and sodium palmitate. Green Chem. 2014;16(4):1921–30.
58. Druel L, Bardl R, Vorwerg W, Budtova T. Starch aerogels: a member of the family of thermal superinsulating materials. Biomacromolecules. 2017;18(12):4232–9.
59. Konstantakos S, Marinopoulou A, Papaemmanouil S, Emmanouilidou M, Karamalaki M, Kolothas E, et al. Preparation of model starch complex hydrogels. Food Hydrocoll. 2019;96:365–72.
60. Alavi F, Ciftci ON. Effect of starch type and chitosan supplementation on physicochemical properties, morphology, and oil structuring capacity of composite starch bioaerogels. Food Hydrocoll. 2023;141:108637.
61. García-González CA, Uy JJ, Alnaief M, Smirnova I. Preparation of tailor-made starch-based aerogel microspheres by the emulsion-gelation method. Carbohydr Polym. 2012;88(4):1378–86.
62. Camani PH, Gonçalo MGM, Barbosa RFS, Rosa DS. Comprehensive insight of crosslinking agent concentration influence on starch-based aerogels porous structure. J Appl Polym Sci. 2021;138(34):e50863.
63. Soleimanpour M, Tamaddon AM, Kadivar M, Abolmaali SS, Shekarchizadeh H. Fabrication of nanostructured mesoporous starch encapsulating soy-derived phytoestrogen (genistein) by well-tuned solvent exchange method. Int J Biol Macromol. 2020;159:1031–47.
64. Zhai Z, Wang S, Xu Y, Zhang L, Wang X, Yu H, et al. Starch-based carbon aerogels prepared by an innovative KOH activation method for supercapacitors. Int J Biol Macromol. 2024;257:128587.
65. Wang Y, Wu K, Xiao M, Riffat SB, Su Y, Jiang F. Thermal conductivity, structure and mechanical properties of konjac glucomannan/starch based aerogel strengthened by wheat straw. Carbohydr Polym. 2018;197:284–91.
66. Ubeyitogullari A, Moreau R, Rose DJ, Zhang J, Ciftci ON. Enhancing the bioaccessibility of phytosterols using nanoporous corn and wheat starch bioaerogels. Eur J Lipid Sci Technol. 2019;121(1):1700229.
67. Ubeyitogullari A, Brahma S, Rose DJ, Ciftci ON. In vitro digestibility of nanoporous wheat starch aerogels. J Agric Food Chem. 2018;66(36):9490–7.
68. Joshi P, Gupta K, Uniyal P, Jana A, Banerjee A, Kumar N, et al. Cassava starch-derived aerogels as biodegradable packaging materials. Mater Chem Phys. 2023;296:127282.
69. Kaur S, Ubeyitogullari A. In vitro digestion of starch and protein aerogels generated from defatted rice bran via supercritical carbon dioxide drying. Food Chem. 2024;455:139833.
70. García-González CA, Smirnova I. Use of supercritical fluid technology for the production of tailor-made aerogel particles for delivery systems. J Supercrit Fluids. 2013;79:152–8.
71. Abhari N, Madadlou A, Dini A. Structure of starch aerogel as affected by crosslinking and feasibility assessment of the aerogel for an anti-fungal volatile release. Food Chem. 2017;221:147–52.

72. García-González CA, Alnaief M, Smirnova I. Polysaccharide-based aerogels - promising biodegradable carriers for drug delivery systems. Carbohydr Polym. 2011;86:1425–38.
73. Alavi F, Ciftci ON. Green and single-step simultaneous composite starch aerogel formation-high bioavailability curcumin particle formation. Int J Biol Macromol. 2024;264(P1):129945.
74. Franco P, Aliakbarian B, Perego P, Reverchon E, De Marco I. Supercritical adsorption of quercetin on aerogels for active packaging applications. Ind Eng Chem Res. 2018;57(44):15105–13.

Chapter 3
Starch Stabilizing Bioactive Compounds

Edith Agama-Acevedo and Luis A. Bello-Perez

3.1 Introduction

The food matrix includes diverse macromolecules, such as proteins, lipids, polysaccharides, and non-starch polysaccharides, as well as bioactive compounds where phenolic compounds are included. There is a rising interest in developing food matrices that provide some health benefits to consumers. The use of raw materials with non-starch polysaccharides, oligosaccharides, resistant starch, dietary fiber components, and bioactive compounds with antioxidant activity has gained interest due to the health-promoting benefits, which include combating oxidative stress associated with some degenerative and chronic diseases. The interactions between two or more of those components of the food matrix affect its physicochemical, functional, sensory, and digestibility characteristics. Some components are present in the raw materials, while others are added in the formulation of the food matrix. Those interactions are produced from the mixing of the ingredients until the packing of the product. Starch is the main component of the diet of human beings; it is present in cereals consumed directly after cooking, such as maize, rice, and oats; legumes as beans, lentils, and chickpeas; tubers and roots, such as potato, sweet potato, and taro; and unripe fruits as plantain. Those starchy sources contain bioactive compounds (polyphenols) that, during processing, interact with the starch and decrease the digestibility of the polysaccharide. In addition, some raw materials used in the preparation of the food matrix are a rich source of polyphenols (orange peels, grape, strawberry). The interactions between starch components, mainly amylose, and polyphenols can be a vehicle to increase the accessibility of bioactive compounds due to the journey through the digestive system, where degradation reduces the amount of polyphenols. Understanding the interactions between

E. Agama-Acevedo · L. A. Bello-Perez (✉)
Instituto Politécnico Nacional, CEPROBI, Yautepec, Mexico
e-mail: labellop@ipn.mx

G. Ayala Valencia (ed.), *Starch: Progress in Food Applications*,
https://doi.org/10.1007/978-3-031-98732-8_3

amylose and polyphenols is important to design a new starch-based release system where polyphenols with antioxidant activity reach the site (tissue or organ) and produce the desired effect (bioavailability).

The present chapter aims to analyze the interactions between bioactive compounds (mainly polyphenols) and starch in model systems and, during food matrix processing, evaluate the release of bioactive compounds from the starch complex and the food matrix, the bioaccesibility and bioavailability of those compounds, and the effect on starch digestibility and glucose transportation toward blood.

3.2 Bioactive Compounds

Bioactive compounds are present in plants and produced during secondary metabolism. They are also called phytochemical compounds and have some effect on human beings; for example, they relieve pain (throat, stomach, grinders, colon) and are used to treat hypertension, hyperlipidemia, obesity, hypercholesterol, and diabetes. Bioactive compounds help to reduce oxidative and inflammatory stress, stimulate the immune system, and modify the digestion of macronutrients (proteins, polysaccharides, lipids). They are present in the indigestible fraction associated with components of the dietary fiber (antioxidant dietary fiber) because they have a prebiotic function on the microbiota given the production of different bioactive compounds with improved functionality during reabsorption to blood. Ethanolic extracts and those of bioactive compounds from plants are consumed as infusions in hot water. Bioactive compounds are present in fruits (grapes, berries, bananas), cereals and pseudocereals (blue and red maize, quinoa, amaranth, sorghum, rice), tubers (sweet potato, potato, taro), legumes (black beans, chickpeas, faba beans), and are consumed directly after cooking or as ingredient in some foods (tortillas, bakery products, snacks, pasta). The bioactive compounds in plants are phenolic acids, flavonoids, polyphenols, and anthocyanidins. Polyphenols are secondary metabolites present in plant foods and include phenolic acids and flavonoids (Fig. 3.1). Phenolic acids are classified as non-flavonoid polyphenol compounds and are derivatives of benzoic acid and cinnamic acid; in the first group (hydroxybenzoic acids) are gallic (GA), vanillic, and syringic acids; and in the second (hydroxycinnamic acids), ferulic, sinapic. Caffeic, and coumaric acids. Flavonoids are considered the major phenolic compounds and impart orange, red, yellow, blue, and purple colors to fruits and leaves; they also show biological functions as signaling during nodulation and protection against ultraviolet radiation. Flavonoids include anthocyanins, flavones, flavonols, isoflavones, flavanones, and flavanols [1]. They present two aromatic rings in their structure linked by a heterocyclic ring with oxygen; they are free monomers but are found as β-glucoside in the plant sources mentioned above.

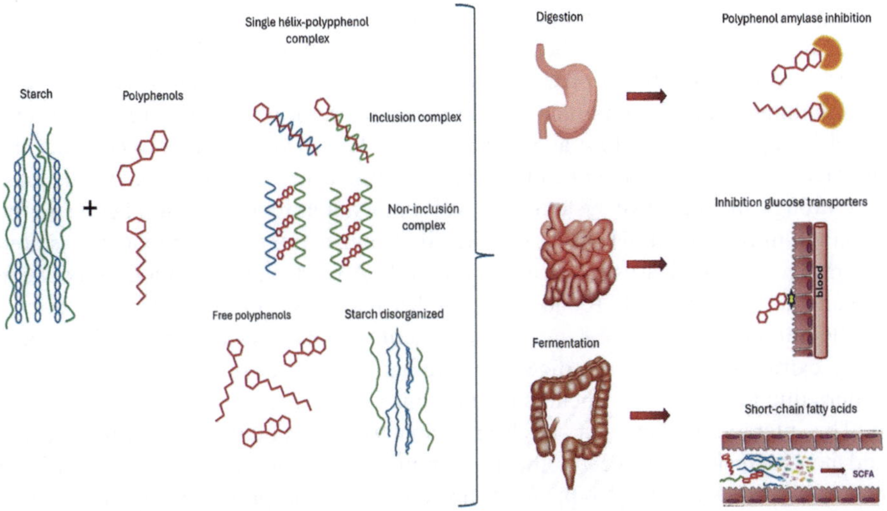

Fig. 3.1 Interaction between starch chains and polyphenol compounds and its impact on human digestion

3.3 Starch–Bioactive Compound Interactions

As mentioned before, starch is present in most of the foods consumed daily and is associated with two main health problems, overweight and obesity, along with related chronic-degenerative diseases such as diabetes, cardiovascular problems, metabolic syndrome, and cancer. Those health issues have also been linked with starch hydrolysis rate, described in the next sections.

The interaction between starch, mainly the linear chains (amylose), and bioactive compounds produces changes in the structure of the polysaccharide that are not recognized by digestive enzymes. For this reason, studies in system models use amylose and high-amylose starch blended with different polyphenols understand the structure, functionality, physicochemical, and digestibility characteristics of the complex. Recently, studies on traditional medicinal plant complexes with starch have suggested that such complexes would be either vehicles for the compounds in extracts or functional ingredients in plant-based foods [2].

The interaction between bioactive compounds and starch aims at: (1) protecting bioactive compounds for and control the release of the guest molecule; (2) decreasing starch hydrolysis rate in the food matrix. The protection of bioactive compounds through the complexation between starch linear chains and the guest molecule is governed by structure, chemical characteristics, and the method to produce the complex where parameters (temperature, mixing, reaction time) are essential [3]. The interactions between starch and bioactive compounds produce a functional ingredient that can be added to ready-to-eat foods or used as a preventative or treatment drug. In a food matrix, they are produced during the processing, when raw materials with starch and bioactive compounds (polyphenols) are mixed and thermally treated, the starch is gelatinized, and the amylose chains are lixiviated from the starch granule toward the continuous phase. Water content, temperature, and process (extrusion, baking,) play an important role in the production of linear chains from both starch components (amylose and amylopectin). It is also important to consider the interactions between starch and bioactive compounds in food ingestion and the journey through the digestive system. It is well known that food digestion begins in the mouth, while the acid pH of the stomach hydrolyzes the main components (starch and proteins) and releases others as polyphenols. A work on the interactions between proteins and polyphenols found this complex may have anti-inflammatory effects on the immune system and some health benefits [4]. The co-digestion of bread and berry extract reduced starch digestion rate and the extent of hydrolyzed starch, a pattern due to starch–polyphenols interactions or amylase inhibition [5].

The interactions between starch and bioactive compounds can form inclusion and non-inclusion complexes, the former producing the crystalline arrangement named V-type where the polyphenol is hosted into a hydrophobic cavity of the linear starch chains via hydrophobic forces and hydrogen bonding. Inclusion complexes between starch and different polyphenols, such as proanthocyanidins [6], isoflavone [7], and curcumin [8], have been reported. The V-type inclusion complex is named VI type and can be considered the classic starch–polyphenol complex; the structural change produced in the polysaccharide modifies some physicochemical characteristics of the starch, like pasting profile [9] and retrogradation properties because the reassociation of the linear chains is prevented [10]. Another association between starch and bioactive compounds named VII is a tail-to-tail arrangement of the two molecules of the bioactive compound into the helix of linear chains that leaves the functional groups at the helical opening [11]. An additional VII type has been proposed as VIIb, where the linear helix can be attached on both ends of the bioactive compounds, producing a more stable complex (high melting temperature). An additional model of the interaction between starch and bioactive compounds is the non-inclusive complex, including CH-π bonds and complex formation by the polyphenol embedded in the hydrophobic cavity of amylose [12]. An extension of this model cannot rule out that the hydrophobic cavity (VI) and the non-inclusive complex (VIIa and VIIb) may be present in a complex formation.

3.4 Evidence of Starch–Phenolic Compound Complex

The complex formation between starch and bioactive compounds requires corroboration. The preparation methods and characterization techniques to evidence the complex formation have been revised [13]. The preparation methods include coprecipitation at high temperatures [8, 14], alcohol [6], dimethyl sulfoxide (DMSO) [15, 16], and thermomechanical methods with high-pressure homogenization [17]. Those studies included characterization by techniques such as X-ray diffraction; small-angle X-ray scattering; differential scanning calorimetry (DSC); nuclear magnetic resonance (NMR); Fourier transform infrared spectroscopy (FTIR); and some microscopy techniques as scanning electron microscopy (SEM), confocal laser scanning microscopy (CLSM), and transmission electron microscopy (TEM).

The complex changes the crystalline structure of starch and some physicochemical and rheological characteristics. Its formation is achieved in two different systems: water and DMSO and guest molecules with different chemical structures [11]. The use of DMSO produces an empty amylose (V-amylose) where the bioactive compound or guest can be inserted [18]. It is well known that starch is insoluble in water at room temperature, so the use of water to produce the complex is carried out at high temperatures (160–180 °C) to dissociate amylose helices and produce a random coil structure of linear chains. After heating, recrystallization of starch components is produced upon cooling with a crystalline structure different (B crystals) from its native counterpart (A crystals). However, the introduction of the guest molecule to the starch solution produces a complex between them and different V-type crystals; in this system, both crystal types (B- and V-type) can be present (Fig. 3.2).

The phase transitions in the starch–polyphenol complex can be determined by DSC and is used to evaluate the formation of the complex. The heating of the sample (complex) in DSC dissociates the complex by producing one or multiple endotherms that indicate the phase transitions at high temperatures. These are higher than those obtained during the phase transition of native starch in excess water, gelatinization, ranging between 80 and 120 °C. The phase transitions of the complex have evidenced interactions between linear chains of starch and polyphenol. Differences in temperatures and enthalpy associated with the phase transitions can be related to the chemical characteristics of the guest molecule and the preparation

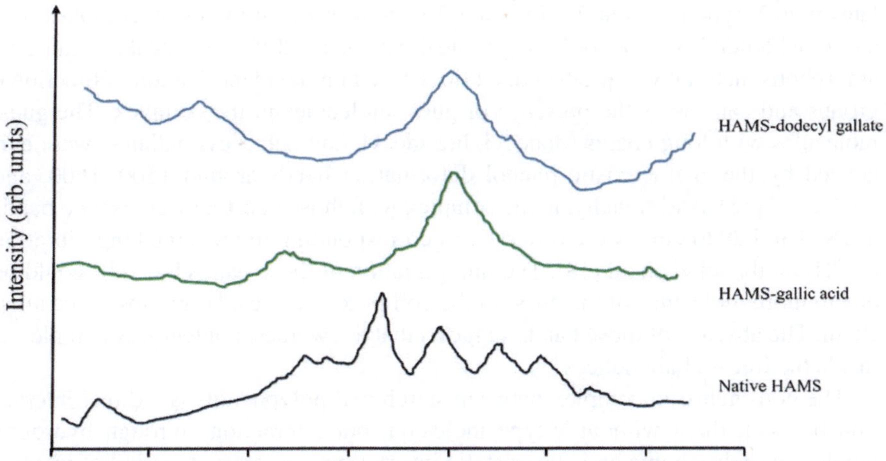

Fig. 3.2 X-ray diffraction spectra of high-amylose maize starch (HAMS) complex with 10% of gallic acid and dodecyl gallate. (Adapted from Zhang et al. [16] and Kong & Ziegler [19])

method of the complex [11]. Guest molecules with various chain lengths, as GA and alkyl gallates, form complexes with amylose and high-amylose maize starch [16]. The formation of the complex has been examined with DSC, X-ray diffraction (XRD), and FTIR spectroscopy. The chain length of the guest molecule determines the type of complex produced, which can be an inclusion complex or a no-inclusion complex, as mentioned above (VIIa and VIIb). The complex produced with GA, butyl gallate, and octyl gallate did not show a phase transition (endotherm) in the DSC study, indicating that an inclusion complex was not formed with both amylose and high-amylose maize starch (Fig. 3.2). The presence of endotherms in the complex longer alkyl chains indicated the formation of the inclusion complex. This pattern results from longer alkyl chains having greater hydrophobicity because they are more prone to staying inside the hydrophobic helix of the linear chains of starch and yielding stronger interactions, observed in the higher enthalpy value. The complex formation between the guest molecule and amylose showed higher interactions than high-amylose maize starch because more energy was necessary to disorganize the complex.

The evidence of the complex is different depending on the method used. The complex formation between linear chains of starch (amylose and long external chains of amylopectin) forms left-handed single helices with a hydrophobic cavity to include the bioactive compound (guest). The guest crystallizes in a V-type structure according to the size and shape of the guest molecule. With aliphatic chains as guests, the linear chain of starch crystallizes into a V-type structure containing six glucose units per helical turn. The V6 structure can be categorized as hydrated V6h or anhydrous V6a; both structures give characteristic X-ray diffraction peaks, the first at 2q 7.5°, 13° and 20°, and the second one at 2q 7.8°, 22° and 24° [11]. XRD showed evidence of a complex with octyl, dodecyl, hexadecyl, and octadecyl gallates with V-type peaks at 2°, 13°, and 20°, indicating an inclusion complex. The amylose blended with GA or butyl gallate did not show diffraction peaks because an amorphous material was produced. FTIR is based on the identification of functional groups and can assess the presence of guest molecules in the complex. The guest molecules with long chains (dodecyl, hexadecyl, and octadecyl gallates) were evidenced by the characteristic phenol deformation bands around 1500, 1600, and 1670 cm^{-1} [15]. Additionally, in the complex with those guest molecules, two bands at 2850 and 2920 cm^{-1} were observed as corresponding to the stretching vibration of CH_2 on the alkyl chain [18]. The interpretation of those bands by FTIR could be due to more molecules of the guest in the complex or more CH_2 groups in the alkyl chain. The absence of those bands suggests that a low guest content was complexed inside the linear chain helices.

The non-inclusion complex between starch and polyphenols is a direct interaction between them without V-type inclusion, but interactions through hydrogen bonds and hydrophobic and electrostatic interactions are present, a kind of interaction that cannot be evidenced by DSC and XRD [1]. Specific interactions between ferulic acid and maize starch do not produce inclusion complex and cannot be detected by DSC and XRD; still, if the interaction is produced, changes in pasting profile, swelling, dynamic rheology, gel texture, and thermal properties (gelatinization and retrogradation) evaluated by DSC, are modified in correlation to the ferulic acid concentration in the complex [20].

The methods used for starch characterization at different levels are considered complementary and provide information on the complex formation between starch chains and bioactive compounds. It is important to consider the chemical structure of the guest molecules to predict the complex formation with a high concentration of the guest molecule.

3.5 Starch Hydrolysis Rate of the Complex

Starch hydrolysis rate gained interest during the 1980s due to the health problems associated with the consumption of digestible carbohydrates. Studies on starch digestion showed that besides digestible starch (rapidly and slowly hydrolyzed), there is a fraction that resists hydrolysis by digestive enzymes in the small intestine. This fraction, resistant starch (RS), is considered part of the dietary fiber. Rapidly digestible starch (RDS) fraction is hydrolyzed during the first 20 min after ingesting starchy food. The consumption of foods with high RDS content is associated with glucose peaks that produce degenerative diseases such as diabetes [21]. Slowly digestible starch (SDS) and RS were named nutraceutical starches because their consumption is linked to some health benefits [22]. The SDS hydrolyzed in the last portion of the small intestine (ileum) and the glucose absorbed in this section of the digestive system produce signals toward the hypothalamus, related to with the production of incretin hormones as GLP1 which produce satiety [23, 24]. RS is classified into five categories [21]; RS5 includes amylose-lipid complexes [25] and, more recently, amylose-polyphenol complexes [26].

Starch hydrolysis can be modulated through interactions between starch and phenolic compounds (inclusion and non-inclusion complex) or between phenolic compounds and digestive enzymes (Table 3.1).

Phenolic compounds may compete with starch for active sites of digestive enzymes (α-amylase and α-glucosidase), resulting in the suppression of the catalytic activity of starch molecules [30, 31]. It has been reported that ferulic acid inhibits both digestive enzymes by two mechanisms (mixed/non-competitive) [32], while two gallates and theaflavin presented an inhibition of the α-amylase mixed-type [33]. Those studies on the inhibition role of the polyphenol on digestive enzymes showed that the inhibitory type strongly depends on the polyphenol structure. Resistance to enzymatic hydrolysis in the intestine of the starch-polyphenol complex will be different if an inclusion complex or non-inclusion complex is produced. Both complexes modified the starch structure that is not recognized by digestive enzymes; however, the non-inclusion complex would release the polyphenol and produce an inhibitory effect on amylases, reducing starch hydrolysis and glucose production [34].

If the release of the polyphenol from the complex is produced in the distal section of the small intestine, the starch would be hydrolyzed to glucose, and the

Table 3.1 Starch digestion fractions, rapidly digestible starch (RDS), slowly digestible starch (SDS), and resistant starch (RS) content of starch-polyphenol complex

Starch	Polyphenol	RDS	SDS	RS	References
High-amylose maize starch (HAMS)	Gallic acid	32	3	64	[16]
	Butyl gallate	37	6	57	
	Octyl gallate	26	16	57	
	Dodecyl gallate	58	13	29	
	Hexadecyl gallate	32	34	34	
	Octadecyl gallate	30	42	29	
Normal maize (5%, dry base)	Caffeic acid (%, dry base)				[27]
	0.0	75	11	14	
	0.6	51	14	35	
	1.0	62	14	24	
Normal corn (10%, dry base)	Isoflavone (%, dry base)				[7]
	0	79	10	11	
	1	68	15	17	
	2	61	17	22	
Debranched cassava starch with pullulanase	Ferulic acid (g/g starch, dry base)				[28]
	Native	Ni	Ni	3	
	0	Ni	Ni	16.0	
	0.02	Ni	Ni	15.9	
	0.05	Ni	Ni	16	
	0.10	Ni	Ni	13.3	
HAMS	Caffeic acid (0.5 g/10 g starch, dry base)				[29]
After gelatinization-retrogradation,		6.8	4.8	88.4	
Starch complex		5.8	5.3	88.9	
Normal					
After gelatinization-retrogradation,		18.8	8.0	73.2	
Starch complex		18.6	6.8	74.6	
Waxy					
After gelatinization-retrogradation,		20.5	10.0	69.6	
Starch complex		20.4	6.8	72.8	

Ni not informed

absorbed glucose in that section is associated with the production of incretin hormones that send signals to the brain, indicating satiety. Although one part of the starch cannot be hydrolyzed due to the possible inhibition of amylases, it is important to analyze the behavior of the non-inclusion complex in this section of the digestive system.

The inclusion complex produces a V-type crystalline structure that resists amylase hydrolysis. It has been mentioned that resistance of the inclusion complex is due to the helix being too big and rigid to fit into the active site of α-amylase. The inclusion complex between starch and polyphenol may be a method to prepare RS and use it as an ingredient in foods that do not require heating or cooking.

Complexes of high amylose maize starch (HAMS) and amylose (AM) with different alkyl gallates have shown different RS content: It was higher in the complex with amylose than in the one with HAMS. The latter favored SDS and RDS production. The increase in the length of the alkyl chain reduced the RDS content since the inclusion complex was produced, modulating the glycemic response [16].

The inclusion complex between amylose and salicylic acid can produce a helix with 6, 7, or 8 glucosyl units per turn: V6, V7, and V8, respectively. It was found that with an excess of the molecule guest, the V8 helix was produced. The hydrolysis rate was different depending on the glucosyl units in the helix [35].

The inclusion complex resists hydrolysis in the small intestine and continues its journey toward the colon because it is a prebiotic. It is recognized as the prebiotic of different types of RS and the products of the fermentation (short-chain fatty acids, SCFAs), mainly butyrate [36].

The starch–polyphenol complex that reaches the colon because the polyphenol is not released during the intestinal phase is fermented there by the microbiota, it is more bioaccessible, and polyphenol is transformed [37, 38]. Both products of the fermentation of the starch-polyphenol complex (SCFAs) and derivative polyphenols are considered healthy molecules. The main health benefits of butyrate are a decrease of inflammatory bowel diseases, immunity, gastrointestinal comfort, and longevity [39, 40], and some strategies to produce butyrate would be the complex production between starch and polyphenols. The relationship between microbiota and health is well known, the modulation of the bacterium species in the colon is due to the components of the diet that reach the colon. Those components, prebiotics, are substrates for the microbiota and stimulate the growth of beneficial bacteria. The starch–polyphenol complex is considered prebiotic and can have different patterns during its metabolism by the microbiota. The type of polyphenol, the bacteria species in the colon of each person, and other components of the diet reaching the colon are factors to consider in studies on the fermentation of starch–polyphenol complexes.

In our opinion, the issue of the complex type (non-inclusion, inclusion, or both) and the remains of the complex or polyphenols released after digestion reaching the colon for fermentation are key to advance the role of the polyphenols on human health.

3.6 Bioaccessibility of Phenolic Compounds

The linear chains of starch from amylose and external long chains of amylopectin can produce complexes with polyphenols. The interactions between these compounds can be considered from diverse points of view: Interactions are produced during food processing (thermal process), where starch and polyphenols are present, or during the co-digestion of meals with the presence of both compounds.

Digesting the starch–polyphenol complex during the gastric and intestinal phases in both matrices reduces the bioaccessibility of polyphenol; that is, the concentration that reaches the blood after intestinal digestion. Polyphenol bioaccessibility is higher in the co-digestion of both compounds than in the food matrix produced by thermal treatment of raw materials with starch and polyphenols. This pattern is because, during co-digestion, the interactions between starch and polyphenols produce non-inclusion complexes that are more easily released as bioaccessibility of polyphenol increases. This pattern was found during the gastric phase; still, no clear increase was observed after the intestinal phase. The polyphenol type influenced bioaccessibility because fewer proanthocyanidins were bioaccessibility in the gastric phase than anthocyanins [5]. The bioaccessibility of polyphenols could be analyzed from diverse angles; first, bioaccessible polyphenols can inhibit α-amylase and α-glucosidase, and the effect of this inhibition reduces starch hydrolysis rate, glucose production, and transportation of glucose toward the blood, which leads to a low glycemic response. Polyphenols can inhibit the glucose transporters that also reduce the glycemic response. The use of polyphenols in both processes decreases the bioaccessibility and bioavailability of polyphenols.

The recovery of polyphenols from cooked porridges after digestion has been studied. It is known that polyphenols are unstable through the intestinal digestion phase due to the pH of the environment (6.5–7.0). Polyphenol recovery was influenced by the size, and GA showing a small molecular size presented a higher recovery (76%) than catechin (43%) and epigallocatechin (16%) [41].

The bioaccessibility of polyphenols in the complex with starch can be analyzed as a protector effect of the starch, mainly in the inclusion complex (encapsulation) for the controled release of polyphenols (bioactive compounds). Polyphenol encapsulation increases stability due to their sensibility to high temperature, pH, and oxidative stress, and they are degraded during processing, storage, and digestion. Those drawbacks can be overcome by encapsulation and ensure their release in the small intestine and absorption into the blood (bioaccessibility). In addition to the protective effect, the digestion of the starch–polyphenol complex could also be delayed, which decreases glucose release toward the blood; this is a positive effect related to diverse metabolic diseases. The slow starch digestion of the complex is a promising issue for the controled and sustained delivery of polyphenols or other guests along the gastrointestinal tract [42]. The study of the encapsulation of diverse food nutrients through a starch complex has been reported [43, 44] using phytochemicals [18, 45–47]. It was found that phytochemicals (bioactive compounds), such as ascorbic acid, retinal, and phytosterols, were complexed with linear chains of starch by inclusion. The solubility of the bioactive compounds in the reaction medium and the structure of the compound can restrict the content of the guest molecule in the complex [46]. The flavor of foods related to the release in the oral cavity during consumption has also been analyzed. A model complex between starch and different flavor compounds showed a range of solubility and hydrophobicity. The complex formation depended on the hydrophobicity and concentration of the compound. Some compounds as limonene did not complex with starch, but in a binary system with menthone or thymol. The authors concluded that the formation was cooperativity-type with a critical ligand concentration to produce a stable complex [48]. However, the release of the compounds was not determined.

Recently, the concentration of total polyphenols was determined after the digestion of encapsulated decoction residues of Roselle with different starchy wall materials. The hypothesis was that, during the extrusion process, anthocyanins of Roselle and linear chains from starch during the extrusion produce a complex. The reduction in total polyphenol content in the extrudates after 360 min of digestion ranged between 33% and 56%; this pattern indicates that the difference is the bioaccessible polyphenol.

A study on starch digestion of polyphenols showed slow hydrolysis. Only 5–10% of the polyphenols present in the complex of the food matrix are released and absorbed in the small intestine. The remaining starch–polyphenol complex enters the colon, which is a substrate for the microbiota. The starch is transformed into SCFAs and polyphenols are transformed into derivatives that are absorbed in the large intestine or discharged with feces [49]. The antioxidant activity of the transformed polyphenols can produce an environments that reduces the free radicals produced or that reach this section of the digestive system, associated with a decrease in colon cancer. Additionally, polyphenols in the colon may promote the growth of probiotic bacteria by regulating the redox state in the large intestine [50].

3.7 Conclusions

Starch–bioactive compound complexes have gained interest due to their health benefits. Bioactive compounds are stabilized through complexation with linear chains of starch by inclusion and non inclusion complex during processing, to produce a food matrix, and co-digestion, thus the bioactive compounds (guest) are protected. The release of bioactive compounds to maintain bioaccessibility is the most important factor in those systems. The challenge is the release of the bioactive compound during intestinal digestion and maintaining its activity. The encapsulation of traditional medicinal plants that present an important number of compounds and the release of those bioactive compounds deserve further investigation. The compounds in those medicinal plant extracts are diverse compounds with different hydrophobicity and chemical structure that can produce inclusion and non-inclusion complexes. In this first step of the encapsulation, it is necessary to know if they produce a complex with starch chains, how the complex is digested in the intestinal phase, which compounds remain in the indigestible fraction that reaches the colon, which products of the fermentation are produced by the microbiota (short-chain fatty acids and polyphenols), and which bacteria species grow.

Acknowledgments The authors gratefully acknowledge the Instituto Politécnico Nacional, CEPROBI for the support.

Conflicts of Interest The authors declare no conflict of interest.

References

1. Bordenave N, Hamaker BR, Ferruzzi MG. Nature and consequences of non-covalent interactions between flavonoids and macronutrients in foods. Food Funct. 2014;5(1):18–34. https://doi.org/10.1039/c3fo60263j.
2. Bello-Perez LA, Santana-Galeana RS, Flores-Silva PC, Tovar J. A cooked pasta matrix high in dietary fiber, polyphenols, and antioxidant capacity using decocted *Hibiscus calyces*. J Cereal Sci. 2024;118:103963.
3. Kong L, Yucel U, Yoksan R, Elias R, Ziegler GR. Characterization of amylose inclusion complex using electron paramagnetic resonance spectroscopy. Food Hydrocoll. 2018;82:82–8.
4. Poojary MM, Hellwing M, Henle T, Lund MN. Covalent bonding between polyphenols and proteins: synthesis of caffeic acid-cysteine and chlorogenic acid-cysteine adducts and their quantification in dairy beverages. Food Chem. 2023;403:134406.
5. Kan L, Oliviero T, Verkerk R, Fogliano V, Capuano E. Interaction of bread and berry polyphenols affects starch digestibility and polyphenols bio-accessibility. J Funct Foods. 2020;68:103924.
6. Amoako DB, Awika JM. Resistant starch formation through intrahelical V-complexes between polymeric proanthocyanidins and amylose. Food Chem. 2019;285:326–33.
7. Wang S, Wu T, Cui W, Liu M, Wu Y, Zhao C, et al. Structure band in vitro digestibility on complex of corn starch with soy isoflavone. Food Sci Nutr. 2020;8:6061–8.
8. Sun L, Warren FJ, Gidley MJ. Soluble polysaccharides reduce binding and inhibitory activity of tea polyphenols against porcine pancreatic α-amylase. Food Hydrocoll. 2018;79:63–70.
9. D'Costa AS, Golding BA, Raval MK, Rolland-Sabaté A, Bordenave N. Probing gallic acid-starch interactions through rapid viscoanalyzer in vitro digestion. Food Res Int. 2023;173:113409.
10. Li C, Luo JX, Zhang CQ, Yu WW. Causal relation among starch chain-length distributions, short-term retrogradation and cooked rice texture. Food Hydrocoll. 2020;108:125577.
11. Kong L, Pérez-Santos DM, Ziegler GR. Effect of guest structure on amylose-guest inclusion complexation. Food Hydrocoll. 2019;97:105188.
12. Guo K, Liu T, Xu A, Zhang L, Bian X, Wei C. Structural and functional properties of starches from root tubers of white, yellow, and purple sweet potatoes. Food Hydrocoll. 2019;89:829–36.
13. Deng N, Deng Z, Tang C, Liu C, Luo S, Chen T, Hu X. Formation, structure and properties of the starch-polyphenol inclusion complex: a review. Trends Food Sci Technol. 2021;12:667–75.
14. Chi C, Li X, Feng T, Zeng X, Chen L, Li L. Improvement in nutritional attributes of rice starch with dodecyl gallate complexation: a molecular dynamic simulation and in vitro study. J Agric Food Chem. 2018;66(35):9282–90.
15. Wang S, Kong L, Zhao Y, Tan L, Zhang J, Du Z, Zhang H. Lipophilization and molecular encapsulation of *p*-coumaric acid by amylose inclusion complex. Food Hydrocoll. 2019;93:270–5.
16. Zhang Y, Gladden I, Guo J, Tan L, Kong L. Enzymatic digestion of amylose and high amylose maize starch inclusion complexes with alkyl gallates. Food Hydrocoll. 2020;108:106009.
17. Zhao B, Wang B, Zheng B, Chen L, Guo Z. Effects and mechanism of high-pressure homogenization on the characterization and digestion behavior of lotus seed starch-green tea polyphenol complexes. J Funct Foods. 2019;57:173–81.
18. Kong L, Ziegler GR. Molecular encapsulation of ascorbyl palmitate in preformed V-type starch and amylose. Carbohydr Polym. 2014;111:256–63.
19. He T, Zhao L, Wang L, Liu L, Liu X, Dhital S, Hu Z, Wang K. Gallic acid forms V-amylose complex structure with starch through hydrophobic interaction. Int J Biol Macromol. 2024;269:129408.
20. Karunaratne R, Zhu F. Physicochemical interactions of maize starch with ferulic acid. Food Chem. 2016;199:372–9.
21. Bello-Pérez LA, Flores-Silva PC, Agama-Acevedo E, Tovar J. Starch digestibility: past, present, and future. J Sci Food Agric. 2020;100(14):5009–16.

22. Magallanes-Cruz PA, Flores-Silva PC, Bello-Pérez LA. Starch structure influences its digestibility: a review. J Food Sci. 2017;82(9):2016–23.
23. Strader AD, Woods SC. Gastrointestinal hormones and food intake. Gastroenterology. 2005;128:175–91.
24. Lee BH, Bello-Pérez LA, Lin AHM, Kim CY, Hamaker BR. Importance of location of digestion and colonic fermentation of starch related to its quality. Cereal Chem. 2013;90(4):335–43.
25. Gutiérrez TJ, Tovar J. Update of the concept of type 5 resistant starch (RS5): self-assembled starch V-type complexes. Trends Food Sci Technol. 2021;109:711–724.26.
26. Romero-Hernández HA, Gutiérrez TJ, Bello-Pérez LA. Can starch-polyphenol V-type complexes be considered as resistant starch? Food Hydrocoll. 2022;124:107226.
27. Zheng Y, Tian J, Kong X, Yang W, Yin X, Xu E, Chen S, Liu D, Ye X. Physicochemical and digestibility characterization of maize starch–caffeic. LWT. 2020;121:108857.
28. Hung VP, Phat NH, Phi NTL. Physicochemical properties and antioxidant capacity of debranched starch–ferulic acid complexes. Starch Starke. 2013;65:382–9.
29. Han M, Bao W, Wu Y, Ouyang J. Insights into the effects of caffeic acid and amylose on in vitro digestibility of maize starch-caffeic acid complex. Int J Biol Macromol. 2020;162:922–30.
30. Li M, Koecher K, Hansen L, Ferruzzi MG. Phenolics from whole grain oat products as modifiers of starch digestion and intestinal glucose transport. J Agric Food Chem. 2017;65(32):6831–9.
31. Li M, Pernell C, Ferruzzi MG. Complexation with phenolic acids affect rheological properties and digestibility of potato starch and maize amylopectin. Food Hydrocoll. 2018;77:843–52.
32. Zheng Y, Tian J, Yang W, Chen S, Liu D, Fang H, et al. Inhibition mechanism of ferulic acid against α-amylase and α-glucosidase. Food Chem. 2020;317:126346.
33. Sun L, Warren FJ, Netzel G, Gidley MJ. 3 or 3′ - Galloyl substitution plays an important role in association of catechins and theaflavins with porcine pancreatic α-amylase: the kinetics of inhibition of α-amylase by tea polyphenols. J Funct Foods. 2016;26:144–56.
34. Giuberti G, Rocchetti G, Lucini L. Interactions between phenolic compounds, amylolytic enzymes and starch: an updated overview. Curr Opin Food Sci. 2020;31:102–13.
35. Guo J, Shi L, Kong L. Structure-digestibility relationships of starch inclusion complex with salicylic acid. Carbohydr Polym. 2023;99:120147.
36. Brouns F, Kettlitz B, Arrigoni E. Resistant starch and "the butyrate revolution". Trends Food Sci Technol. 2002;3(8):251–61. https://doi.org/10.1016/s0924-2244(02)00131-0.
37. Rocchetti G, Chiodelli G, Giuberti G, Lucini L. Bioaccessibility of phenolic compounds following in vitro large intestine fermentation of nuts for human consumption. Food Chem. 2018;245:633–40.
38. Rocchetti G, Senizza A, Gallo A, Lucini L, Giuberti G, Patrone V. *In vitro* large intestine fermentation of gluten-free rice cookies containing alfalfa seed (Medicago sativa L.) flour: a combined metagenomic/metabolomic approach. Food Res Int. 2019;120:312–21.
39. Louis P, Flint HJ. Formation of propionate and butyrate by the human colonic microbiota. Environ Microbiol. 2017;19:29–41.
40. Teichmann J, Cockburn DW. In vitro fermentation reveals changes in butyrate production dependent on resistant starch source and microbiome composition. Front Microbiol. 2021;12:640253.
41. Ferruzzi MG, Hamaker BR, Bordenave N. Phenolic compounds are less degraded in presence of starch than in presence of proteins through processing in model porridges. Food Chem. 2020;309:125769.
42. Shi L, Zhou J, Guo J, Gladden I, Kong L. Starch inclusion complex for the encapsulation and controlled release of bioactive guest compounds. Carbohydr Polym. 2021;2021(274):118596.
43. Lesmes U, Barchechath J, Shimoni E. Continuous dual feed homogenization for the production of starch inclusion complexos for controlled release of nutrients. Innov Food Sci Emerg Technol. 2008;9:507–15.
44. Lesmes U, Cohen SH, Shener Y, Shimoni E. Effects of long chains fatty acid unsaturation on the structure and controlled release properties of amylose complexes. Food Hydrocoll. 2009;23:667–75.

45. Cohen R, Orlova Y, Kovalev M, Ungar Y, Shimoni E. Structural and functional properties of amylose complex with genistein. J Agric Food Chem. 2008;56:4212–8.
46. Lay Ma UV, Floros JD, Ziegler GR. Formation of inclusion complexes of starch with fatty acid esters of bioactive compounds. Carbohydr Polym. 2011;83:1869–78.
47. Kong L, Yucel U, Yoksan R, Elias RJ, Ziegler GR. Characterization of amylose inclusion complexes using electron paramagnetic resonance spectroscopy. Food Hydrocoll. 2018;82:82–8.
48. Yeo L, Thompson DB, Peterson DG. Inclusion complexation of flavour compounds by dispersed high-amylose maize starch (HAMS) in an aqueous model system. Food Chem. 2016;199:393–400.
49. Cardona F, Andres-Lacueva C, Tulipani S, Tinahones FJ, Queipo-Ortuno MI. Benefits of polyphenols on gut microbiota and implications in human health. J Nutr Biochem. 2013;24:1415–22.
50. Yang C, Deng Q, Xu J, Wang X, Hu C, Tang H, et al. Sinapic acid and resveratrol alleviate oxidative stress with modulation of gut microbiota in high-fat diet-fed rats. Food Res Int. 2019;116:1202–11.

Chapter 4
Starch Stabilizing Proteins

Wilson Daniel Caicedo Chacon, Guilherme Dallarmi Sorita, Valeria Cedeno Fierro, and Juan Felipe Grisales Mejía

4.1 Introduction

Proteins are essential macromolecules in all living organisms, performing structural, enzymatic, regulatory, and transport functions. These biomolecules are composed of chains of amino acids, and their three-dimensional structure is fundamental to their biological activity [1, 2]. A protein's specific conformation depends on weak interactions, such as hydrogen bonds, hydrophobic interactions, disulfide bonds, and van der Waals forces, which provide stability and functionality in a specific environment [1]. Protein function is crucial in food, pharmaceutical, and biomedical applications. In the food sector, proteins are key in texture, flavor, and product stability. For example, casein and whey proteins are commonly used as stabilizing and texturizing agents in dairy products and processed foods [3, 4]. The pharmaceutical industry also relies on proteins, such as enzymes and antibodies, to develop targeted treatments, where stability is critical to product efficacy and shelf life [5, 6].

W. D. C. Chacon (✉)
Ingeniería de Alimentos, Fundación Universitaria Agraria de Colombia, Bogotá, Colombia

Department of Chemical and Food Engineering, Federal University of Santa Catarina, Florianópolis, SC, Brazil

G. D. Sorita
Department of Chemical and Food Engineering, Federal University of Santa Catarina, Florianópolis, SC, Brazil

V. C. Fierro
Facultad de Ingeniería de Administración, Universidad Nacional de Colombia, Palmira, Colombia

J. F. G. Mejía
Chemistry Department, Laboratory of Advanced Analytical Techniques in Natural Products (LATNAP), Universidad de los Andes, Bogotá, Colombia

G. Ayala Valencia (ed.), *Starch: Progress in Food Applications*,
https://doi.org/10.1007/978-3-031-98732-8_4

Despite their importance, proteins are highly sensitive to external factors that can affect their structure and functionality. Changes in pH, extreme temperatures, the presence of salts and solvents, as well as handling and storage processes, can denature proteins—altering their three-dimensional structure and causing them to lose activity [7–9]. This phenomenon, known as destabilization, poses a significant problem in applications where protein stability is essential. A common issue is the tendency of proteins to aggregate or precipitate under unsuitable conditions, affecting their solubility and activity [10]. These issues limit their use in commercial and scientific applications, as denaturation or aggregation can drastically reduce the shelf life and functionality of products containing them. For example, in the food industry, proteins that denature during processing may lose properties like water retention capacity or stability in emulsions, impacting the quality of the final product [11, 12]. In pharmaceutical applications, stability issues can lead to losses in the effectiveness of therapeutic proteins [13, 14].

To counteract these challenges, various stabilization strategies have been developed to help maintain protein functionality under adverse conditions. These strategies include the addition of excipients, protein structure modification through genetic engineering, and the use of stabilizing materials, such as polymers and carbohydrates [15]. Each of these techniques has its advantages and limitations, and the appropriate method depends largely on the specific characteristics of the protein and the environment in which it will be used. Among the stabilizers used, polysaccharides have proven especially effective due to their ability to form protective matrices that interact with proteins, preventing denaturation and aggregation [15]. These compounds can create a network that protects the protein by encapsulating it or by interacting with it specifically, contributing to the stability of its structure [16, 17]. In this context, starch, a widely available and economical polysaccharide, has emerged as a promising stabilizer in the food industry and other fields.

Starch is a polysaccharide composed of glucose units, abundant and low-cost, widely used in the food industry due to its functional versatility. Its structure consists of two types of molecules: amylose, which is linear, and amylopectin, which is branched [18]. These structural characteristics enable starch to interact with proteins through various mechanisms, making it an excellent stabilizing agent [18]. The use of starch as a protein stabilizer has been studied in various applications due to its ability to improve the thermal and pH stability of proteins and to reduce their tendency to aggregate [18–20]. In aqueous solutions, starch can form gels and complexes that encapsulate proteins, protecting them from destabilizing factors [21–23]. For instance, in food emulsions, modified starch can form a protective layer around proteins, improving the stability of the emulsion and preventing protein denaturation under storage conditions [21, 24]. Moreover, starch can be chemically modified to optimize its stabilizing capacity under different conditions. Modifications such as esterification or oxidation allow the properties of starch to be adjusted to enhance compatibility with specific proteins, facilitating its use in various applications [25, 26]. These properties make starch a valuable tool for protein stabilization in products where durability and functionality are crucial. This chapter systematically analyzed the stabilization of proteins using starch as a protective agent, as well as summarized the structural properties of proteins and the external factors that lead to destabilization, including denaturation and aggregation.

4.2 Structure, Types, and Function of Proteins

Proteins are complex macromolecules essential for life, serving as the structural and functional building blocks of cells and organisms. They are composed of long chains of amino acids linked by peptide bonds in a specific sequence that determines the protein's unique shape and function [1, 2]. Each protein folds into a specific three-dimensional structure, guided by interactions between amino acids, which is critical for its activity [1, 2]. However, proteins can be highly sensitive to environmental factors, which can cause them to lose their structure and function. This destabilization process, known as denaturation, has significant implications in fields such as food science, biotechnology, and pharmaceuticals [7–9].

4.2.1 The Structure and Function of Proteins

Proteins perform an incredible variety of functions in biological systems. They serve as enzymes, which catalyze biochemical reactions; as antibodies, which help in immune defense; and as hormones and signaling molecules that regulate various biological processes [27, 28]. Additionally, proteins provide structural support, as seen with collagen in connective tissues and keratin in hair and nails [29]. The functional diversity of proteins arises from the variability in their amino acid sequences and the specific folding patterns they adopt [1, 2]. Proteins comprise 20 different amino acids, each with unique side chains that influence the molecule's overall behavior. The primary structure of a protein is its amino acid sequence, while its secondary structure includes repetitive folding patterns, such as alpha helices and beta sheets, stabilized by hydrogen bonds [1]. The tertiary structure refers to the overall three-dimensional shape of the protein, formed by various interactions among side chains, including hydrophobic interactions, hydrogen bonds, ionic interactions, and disulfide bridges [30, 31]. In some cases, proteins may also assemble into a quaternary structure, where multiple protein subunits interact to form a functional complex [32]. A protein's specific folding and structural integrity are essential for its activity. For example, an enzyme's active site, the region where substrates bind, and reactions occur—depends on precise folding [33]. If a protein loses its shape, it also loses its functional capabilities, which can be detrimental in biological and industrial applications [34].

4.2.2 Factors Causing Protein Destabilization

Proteins are sensitive to various environmental conditions, such as temperature, pH, and the presence of chemicals or mechanical stress, which can disrupt their stability [35–39]. Protein destabilization, or denaturation, involves the unfolding or misfolding of a protein's structure, leading to loss of function (Fig. 4.1). Examining the physical and chemical forces that hold their structure together is essential to understanding why proteins are prone to destabilization.

Temperature High temperatures are a common cause of protein denaturation. Heat increases the kinetic energy of molecules, causing the vibrations within a protein to intensify. When the kinetic energy becomes too high, the weak interactions that stabilize the protein's structure, such as hydrogen bonds and hydrophobic interactions, can break [37, 38]. For example, the unfolding of egg white proteins during cooking is a classic example of thermal denaturation [40]. The increased kinetic energy disrupts the interactions holding the proteins in their native state, forming a network that solidifies as it cools [41, 42]. Maintaining proteins at low temperatures is essential in many industrial and food applications to preserve their structure and functionality [43, 44].

pH Changes Proteins are also highly sensitive to pH levels, as the ionization of amino acid side chains depends on the surrounding pH [35–37]. Amino acids like aspartic acid, glutamic acid, lysine, and arginine have side chains that can gain or lose protons, changing their charge based on the pH of the environment [45, 46]. When the pH deviates significantly from a protein's optimal range, it can disrupt the ionic interactions and hydrogen bonds that stabilize its structure [35–37]. For instance, proteins may unfold in highly acidic or basic environments due to repul-

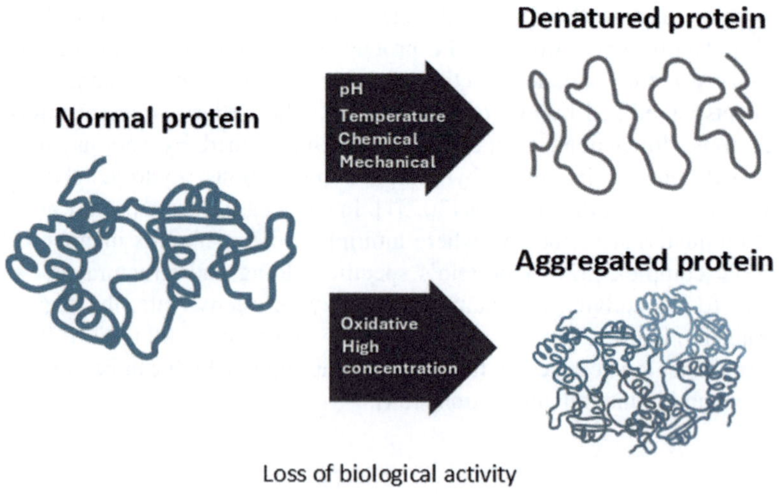

Fig. 4.1 Graphical representation of protein denaturation and aggregation by various factors

sive forces between similarly charged groups, leading to destabilization. This pH sensitivity poses challenges in industries like food processing, where products undergo pH changes during preservation and storage.

Chemical Denaturants Certain chemicals, known as denaturants, can directly interfere with the stability of proteins. Urea and guanidinium chloride, for example, are known to disrupt hydrogen bonds and weaken hydrophobic interactions within proteins [38]. These compounds penetrate the protein's structure, destabilizing the hydrogen bonding network and causing the protein to unfold [47]. Organic solvents, such as ethanol and acetone, can also cause denaturation by disrupting hydrophobic interactions and altering the protein's solubility [48]. Understanding the effects of chemical denaturants is crucial in protein purification and stabilization processes, where the use of certain chemicals can influence the protein's final form [49].

Mechanical Stress Proteins can also undergo structural changes when exposed to mechanical forces [39]. Shear stress, commonly encountered during stirring, mixing, or high-pressure homogenization, can induce partial unfolding or aggregation in proteins [50–52]. For example, in the dairy industry, proteins in milk and cream can denature when subjected to high shear forces, impacting the texture and stability of products [53]. Mechanical denaturation can be particularly challenging in formulations containing high concentrations of protein, where aggregates formed by unfolded proteins may affect the quality and appearance of the final product [54].

Oxidative Stress Oxidative agents can lead to protein destabilization by modifying specific amino acid residues, such as methionine and cysteine [55, 56]. These residues contain sulfur atoms that are particularly susceptible to oxidation, leading to changes in the protein's structure [57]. Oxidative modifications can alter the protein's charge and hydrophobicity, causing it to lose its native conformation and, consequently, its functionality [58, 59]. In biological systems, oxidative stress is a major factor in the degradation of proteins, and similar issues arise in industrial applications where proteins are exposed to oxygen or reactive chemicals [58, 59].

4.2.3 Protein Aggregation and Loss of Function

When proteins denature, they are prone to aggregation, where unfolded or partially folded proteins stick together to form larger complexes [4]. Aggregation is often irreversible, and the resulting complexes can precipitate out of solution [4]. This phenomenon is problematic in various fields. In food science, for example, protein aggregation can impact the texture and stability of products, while in pharmaceuticals, protein aggregation can reduce the efficacy of therapeutic proteins, such as monoclonal antibodies [54, 60]. Aggregation occurs because the hydrophobic regions of unfolded proteins become exposed, and these regions tend to associate with each other to minimize contact with the aqueous environment [10]. In some cases, denatured proteins can partially refold into non-native structures that expose hydrophobic residues, further increasing aggregation [61]. Preventing aggregation is a major challenge in protein stabilization, as it often requires specific excipients or controlled environmental conditions [10, 62].

4.2.4 Strategies for Protein Stabilization

To maintain protein functionality in industrial applications, several strategies are employed to stabilize their structure. Additives such as sugars, polyols, and certain polymers can protect proteins by creating a protective environment around them, reducing the likelihood of unfolding [63, 64]. Additionally, encapsulation techniques using polysaccharides, such as starch, can form a matrix around proteins to guard against destabilizing agents [65]. For example, starch-based matrices have been used to stabilize proteins under varying pH and temperature conditions, particularly in food products where these factors fluctuate [65–68]. Understanding the molecular basis of protein stability and destabilization is critical for optimizing protein-based applications. As science advances, new stabilization methods will emerge, enhancing protein-containing products' life, safety, and efficacy across various industries.

4.3 Mechanisms of Protein Stabilization with Starch

4.3.1 Protein-Starch Interactions

Protein-starch interactions are fundamental in stabilizing proteins, enhancing their functionality, and improving the textural properties of food systems. These interactions occur through a combination of non-covalent bonds, including hydrogen bonding, hydrophobic interactions, and electrostatic forces. Hydrogen bonds form between the hydroxyl groups of starch and the functional groups of proteins, such as amino or carboxyl groups, stabilizing the tertiary and quaternary protein structures. Hydrophobic interactions, on the other hand, reduce protein aggregation by shielding the hydrophobic regions of proteins from water exposure. Electrostatic interactions, dependent on the charge states of both molecules, modulate the affinity and strength of these bonds, significantly influenced by environmental factors like pH and ionic strength [69].

The work by Peng et al. [70] provides critical insights into these mechanisms. The authors describe how gluten networks, particularly those with high-molecular-weight glutenin subunits (HMW-GSs), bind to starch granules through hydrogen bonds, hydrophobic forces, and glycosidic linkages (Fig. 4.2). These interactions inhibit starch gelatinization and digestion by limiting water absorption and enzymatic hydrolysis. For example, amylose and amylopectin fragments from

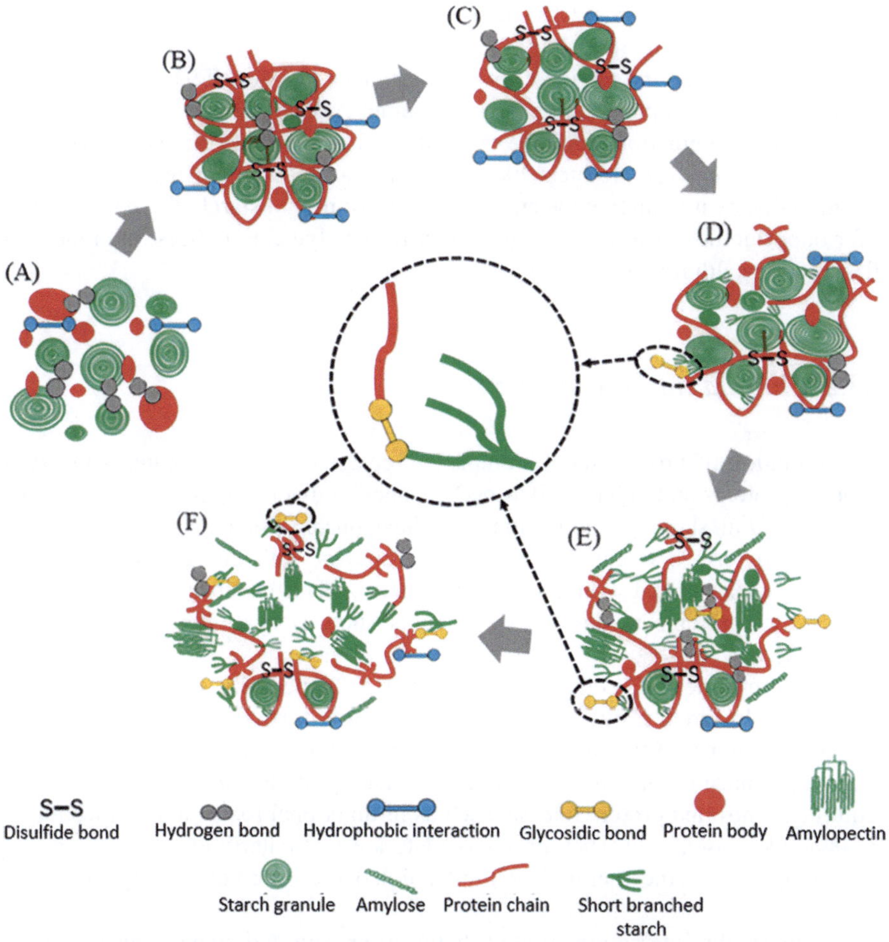

Fig. 4.2 The simplified model reflecting the behavior of protein and starch during the heating and stirring process. (A), (B), (C), (D), (E), and (F) were respectively on behalf of native flour, C1, C2, C3, C4 and C5 in the Mixolab-curve. (Adapted from Peng et al. [70])

gelatinized starch granules form bonds with monomeric proteins, such as globulins and gliadins, creating a dense matrix. This matrix not only stabilizes proteins but also reinforces the structural integrity of the food product during thermal and mechanical processing.

Structural properties of the starch also play a vital role in determining the strength and nature of these interactions. Amylose-rich starches tend to form stronger complexes due to their linear chains, which facilitate close packing and multiple points of interaction with proteins. This is especially critical in applications requiring enhanced thermal and mechanical stability. Peng et al. highlight that stronger protein-starch interactions were observed in wheat systems with 5 + 10 HMW-GSs, compared to weaker interactions in 2 + 12 systems. This difference is attributed to the increased ability of 5 + 10 systems to form robust networks through disulfide bonds and hydrophobic interactions [70].

These molecular interactions not only enhance the functional and structural properties of the complexes but also impact practical applications. In food systems, protein-starch interactions improve viscosity, gel strength, and moisture retention, which are critical for products like bread, pasta, and gluten-free formulations. Additionally, by modulating the enzymatic digestibility of starch, these interactions offer opportunities to design products with lower glycemic indices, aligning with dietary and health trends [71, 72].

4.3.2 Formation of Complexes

The formation of protein-starch complexes represents a sophisticated strategy to stabilize proteins and enhance their performance in diverse applications. This process begins with starch gelatinization under heat, during which its granular structure dissolves to form a viscous paste. Proteins interact with this gelatinized starch, becoming encapsulated within a protective matrix. This encapsulation reduces the exposure of proteins to destabilizing agents such as heat, light, and oxidative stress [73].

The effectiveness of complex formation is influenced by starch composition. Amylose-rich starches, due to their linear structure, form denser and more stable complexes compared to amylopectin-dominant starches. These complexes exhibit higher resistance to enzymatic degradation, making them suitable for applications requiring controlled protein release, such as in functional foods and nutraceuticals. Proteins like casein and whey protein isolate are particularly effective at forming such complexes, as they naturally exhibit a high degree of compatibility with starch [74, 75].

Moreover, the interaction strength between protein and starch depends on the environmental conditions, including temperature, pH, and ionic strength. For example, in acidic conditions, proteins and starches are more likely to form tightly bound complexes due to enhanced electrostatic interactions. Techniques such as confocal laser scanning microscopy (CLSM) have revealed how proteins are embedded within starch matrices, forming distinct layers that act as barriers against external stresses [75].

In food systems, these complexes improve textural properties such as gel strength, elasticity, and viscosity. They are instrumental in creating desirable textures in products like yogurt, pudding, and gluten-free baked goods. Furthermore, starch-protein complexes play a pivotal role in reducing retrogradation, which can lead to staleness in starchy foods. Their application extends to pharmaceutical formulations, where they are used to protect bioactive proteins during storage and digestion [76, 77].

4.3.3 Reduction of Aggregation

Aggregation of proteins is a major challenge, particularly in food and pharmaceutical formulations, where stability and functionality are paramount. Starch mitigates protein aggregation by acting as a physical and molecular barrier. During complex formation, starch molecules surround individual protein molecules, preventing their hydrophobic regions from coming into contact and reducing the likelihood of insoluble aggregate formation. This effect is especially critical during thermal processing, where proteins are prone to denaturation and subsequent aggregation (Scott).

Hydrogen bonds between starch and proteins stabilize protein structures, ensuring that their functional groups remain intact. These bonds reduce the conformational flexibility of proteins, further minimizing aggregation. Additionally, the gelatinized starch matrix acts as a medium that disperses protein molecules evenly, reducing the chances of intermolecular interactions that lead to aggregation [78].

Interestingly, cooperative effects between starch and proteins have also been observed, where each component enhances the stability of the other (Fig. 4.3). Starch plays a protective role by reducing protein denaturation and aggregation under stress conditions such as heating or freezing. Conversely, proteins contribute to reducing the retrogradation of starch by interacting with amylose and amylopectin chains, thus inhibiting their reassociation into crystalline forms. This mutual stabilization results in improved textural and functional properties of the combined matrix, making these complexes highly beneficial in food systems [69].

Experimental studies using whey protein and corn starch have demonstrated that higher starch concentrations correlate with reduced protein aggregation during heating. The presence of starch not only prevents aggregation but also enhances the solubility of proteins in aqueous systems, making them more functional in various applications. This property is particularly beneficial in creating high-protein beverages and nutritional supplements [79, 80].

Furthermore, starch-protein systems exhibit enhanced storage stability, as the starch matrix prevents protein molecules from interacting over time. This is especially valuable in frozen or refrigerated products, where aggregation often occurs during freeze-thaw cycles. By reducing aggregation, starch not only preserves the functional properties of proteins but also extends the shelf life of the final product [81].

4.4 Protein-Starch Stability Under Variable Conditions

Research on protein-starch complexes has gained importance due to their ability to influence starch's functional, structural, and digestibility properties. These interactions are fundamental for the development of functional foods, including those with a low glycemic index, and they also have applications as wall materials for controlled-release drugs.

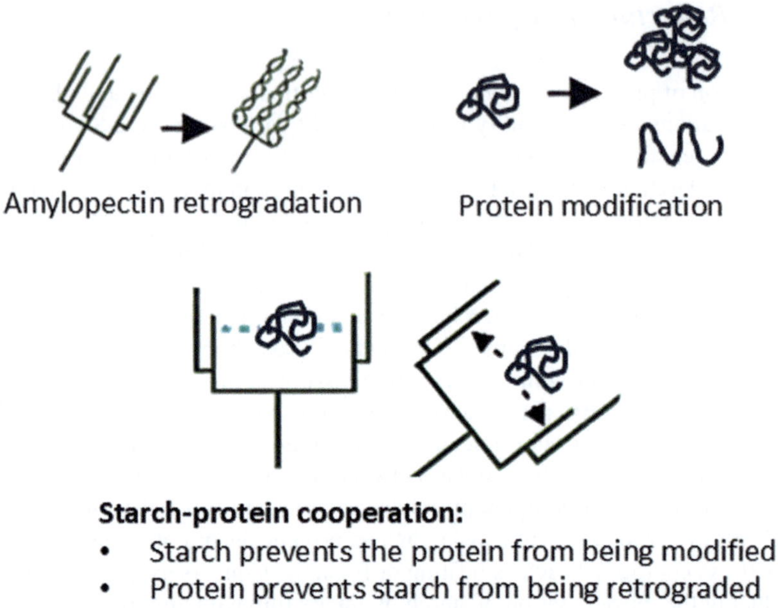

Fig. 4.3 Starch-protein interaction to prevent modifications. (Adapted from Scott and Awika [69])

4.4.1 Influence of Temperature on Complex Stability

4.4.1.1 Gelatinization and Denaturation

Temperature is a key factor in the transformation and functionality of protein-starch complexes, affecting their molecular structure and physicochemical properties in ways that impact their potential uses. At moderate temperatures (60–80 °C), starch undergoes gelatinization, during which amylose and amylopectin molecules absorb water, lose their crystalline organization, and form a viscous matrix. Meanwhile, proteins undergo thermal denaturation, breaking non-covalent bonds and structural changes that expose additional reactive sites for interaction with starch. According to Lu et al. [82], this process can facilitate the formation of more homogeneous and thermodynamically stable structures, improving the complex's susceptibility to enzymatic hydrolysis during digestion.

At higher temperatures (90–120 °C), changes in both molecules may lead to bond formation, resulting in complexes with reduced solubility and digestibility. For instance, thermal extrusion treatments enhance the complex's resistance to enzymatic hydrolysis, which is beneficial for applications such as slow-release functional foods [83].

4.4.1.2 Functional Properties and Molecular Arrangement

Temperature-induced changes affect functional properties such as water retention and emulsifying capacity, critical in food products requiring a balance between structural stability and digestibility. According to Xie et al. [83], thermal treatment induces a reorganization of amylose and protein chains, enhancing their cohesion and reducing deformation during cooling. These properties are vital for food applications where textural stability is essential.

Protein denaturation due to thermal phenomena can also alter certain techno-functional properties, such as water and oil retention capacities, which are fundamental for complexes during food processing. Structural changes modify the complex's physicochemical properties and impact its behavior in industrial applications. Allen et al. [84] observed that a covalent complex formed between amylose and whey protein in an extruded product reduced the water retention capacity of starch and proteins. Similarly, Chu et al. [85] identified that gels obtained with isolated soy protein and corn starch significantly improved their strength and water retention capacity. This improvement was attributed to protein denaturation, which exposes hydrophobic groups that establish similar bonds with starch, strengthening hydrophobic interactions in the gel system.

Temperature also affects starch crystallization and the rheological properties of the system. For example, in ternary starch-fatty acid-protein systems, X-ray diffraction (XRD) analyses reveal the formation of V-type crystals, indicating greater molecular organization. These changes correlate with increased viscosity during cooling due to starch chain reorganization in the presence of proteins, such as gluten, and reduced swelling and solubility capacities. These complexes exhibit lower digestibility due to their higher crystallinity and structural order, acting as physical barriers to digestive enzymes and increasing the proportion of resistant starch [86].

4.4.2 Influence of pH on Complex Stability

pH regulates the electrostatic properties and solubility of components in protein-starch complexes. At a pH near the isoelectric point of proteins (typically between 4 and 5), they tend to aggregate due to reduced electrostatic repulsion. In general, acidic environments ($pH < 3$) cause medium ionization, leading to protein protonation and acquiring a positive charge, destabilizing complexes due to repulsion between negatively charged starch chains. According to Bravo-Núñez et al. [71], acidic pH decreases the solubility of animal proteins, especially near the isoelectric point (pI), favoring aggregate formation and reducing water interaction. Dong and Cui [87] observed similar behavior in complexes formed by potato starch and soy protein. Conversely, plant proteins exhibit greater solubility near their pI. Under alkaline conditions ($pH > 8$), proteins partially denature, exposing hydrophobic groups that promote non-covalent interactions with starch, reducing solubility. This phenomenon has been exploited in food applications to modify texture.

4.4.2.1 Functional Properties and Molecular Arrangement

pH significantly affects the functional properties of starch-protein complexes. Water-binding capacity (WBC) tends to increase at higher pH values, particularly in plant proteins like pea protein. In contrast, animal proteins, such as whey and albumin, exhibit nearly zero WBC in most cases, reflecting their reduced solubility near their isoelectric points [71].

The emulsifying capacity and stability of complexes vary with pH, reaching a minimum at pH 4.0, accompanied by a considerable increase in these properties under more acidic conditions [87].

In terms of viscosity, Bravo-Núñez et al. [71] found that adding proteins to starch tends to reduce peak viscosity due to starch content dilution and interactions limiting starch granule swelling. This effect intensifies under acidic pH conditions, where hydrolysis of starch's amorphous regions generates low-molecular-weight dextrins, reducing gelatinization capacity.

pH also plays a crucial role in the cohesion and organization of gel structures formed by protein-starch complexes. At acidic pH, animal proteins often form visible aggregates, whereas plant proteins, such as pea protein, generate more integrated structures with starch chains. This behavior aligns with studies indicating that reduced protein solubility near their isoelectric points (pI) favors aggregate formation [71, 72].

4.4.3 Stability During In-Vitro Digestion

The protein-starch complex has gained importance in research due to its ability to modify starch digestibility and its potential in developing low-glycemic-index foods. The interaction between proteins, whether endogenous or exogenous, and starch plays a decisive role in its enzymatic hydrolysis. Proteins can form physical barriers by adsorbing or embedding themselves into starch granules, limiting the access of amylases and reducing digestibility. Additionally, these proteins can directly interact with starch molecules, decreasing their enzymatic degradation and attenuating the glycemic response. Furthermore, certain proteins directly inhibit the enzymes responsible for starch hydrolysis, further slowing its digestion [2].

According to the findings of Chen et al. [2], soy peptides enhance starch stability against enzymatic digestion in vitro. The addition of these peptides, especially in combination with heat-moisture treatment (HMT), significantly reduced the content of rapidly digestible starch (RDS) and increased the resistant starch (RS) content. This was attributed to molecular interactions between peptides and starch, leading to more rigid structures that were less accessible to hydrolytic enzymes. This effect was more pronounced in potato starch than corn starch, likely due to differences in their crystalline structures (B-type vs. A-type, respectively). Moreover, HMT-treated complexes exhibited greater resistance to digestion, suggesting that such treatment promotes the structural stability of starch-protein interactions.

Lin et al. [88] explored the effect of rice protein hydrolysates (RPH) on the digestibility of wheat starch during in vitro digestion. RPH demonstrated significant capacity to reduce wheat starch digestibility, which could be attributed to three main mechanisms. First, they acted as physical barriers by adsorbing or embedding themselves onto the starch surface, hindering enzymatic access through hydrophobic interactions and hydrogen bonding. Second, RPH non-covalently bound to starch through hydrogen bonds and hydrophobic interactions, decreasing its availability as an enzymatic substrate. Third, RPH interacted with amylase molecules in a non-catalytic manner, competitively inhibiting enzymatic action. As a result, there was a reduction in RDS content and an increase in RS, with RS reaching up to 46.7% with the incorporation of 12% RPH4. This increase in RS content was attributed to the ability of RPH to inhibit starch recrystallization and limit access by digestive enzymes.

Meanwhile, Lu et al. [89] demonstrated that complexes formed by both endogenous and exogenous rice proteins and rice starch are essential for reducing starch digestibility in vitro. The incorporation of proteins or their hydrolysates significantly decreased the starch digestion rate, as evidenced by lower reaction rate constants (k) compared to unmodified starch. This effect is related to the role of proteins as physical barriers, hindering the interaction between digestive enzymes and starch particles. FTIR and XRD analyses identified the formation of hydrogen bonds between protein hydrolysates and starch, reinforcing the ordered structure of starch and increasing its resistance to enzymatic hydrolysis. Protein hydrolysates showed a more pronounced effect than intact proteins, likely due to the presence of peptides and free amino acids generated during hydrolysis, which further contribute to the structural stability of starch. This promotes a higher content of RS and slow-digesting starch (SDS).

4.5 Practical Applications of Protein-Starch Complexes

Protein-starch complexes have attracted significant scientific attention due to their promising impact on food properties and functionalities. Recent research has explored their formation, behavior, and applications in food systems, revealing innovative ways to improve texture, stability, digestibility, and nutritional value.

One key area of study is how protein-starch interactions affect starch retrogradation, a process central to food stability and texture. In a review article, Scott and Awika [90] described that these interactions are influenced by factors such as the type of protein, starch-to-protein ratios, and processing conditions. For instance, proteins with high water retention capacity, such as soy proteins, reduce water mobility within the matrix, limiting retrogradation and enhancing storage stability. In contrast, less soluble proteins tend to encourage retrogradation by forming covalent bonds and promoting crystalline reorganization, which can impact products like bread, noodles, and gels. This knowledge is crucial for optimizing various foods' shelf life, texture, and glycemic index.

Beyond retrogradation, protein-starch complexes have shown potential in developing meat analogs. Dobson et al. [91] examined matrices made from modified corn starch and pea protein. By combining thermally inhibited starch and pea protein in a 70:30 ratio, they achieved synergistic interactions that enhanced elasticity, chewiness, and structural strength. Spectroscopic and mechanical analyses confirmed these improvements, highlighting the potential of such complexes for creating sustainable, functional alternatives in the food industry.

Another promising application involves improving dough properties. Shi et al. [92] evaluated extruded complexes of yam starch and soy protein isolate in dough formulations. These complexes stabilized the dough structure by reducing free water, promoting a uniform network, and improving functionality. While slight decreases in hardness and elasticity were observed, key mechanical attributes remained intact, showcasing their value in baked goods.

In baking, protein-starch complexes have also proven effective in enhancing bread quality. Zheng et al. [93] investigated substituting whole-grain flour with exogenous proteins in sugar-free and fat-free bread. Proteins like gluten and barley protein improved the dough's rheological, thermal, and structural properties. Notably, barley protein formed stable networks that improved bread texture and lowered starch digestibility, reducing the bread's glycemic index. This positions barley protein as a promising ingredient for functional baking applications.

Protein-starch complexes have additional applications in creating functional powders and beverages. Liu et al. [94] developed powders from oats and barley using milling and extrusion. The resulting complexes protected starch from gelatinization, reducing its digestibility. These effects decreased postprandial glycemic response and promoted carbohydrate fermentation in the colon, contributing to gut health. Such powders are particularly suitable for functional beverages with enhanced nutritional profiles.

The functional versatility of protein-starch complexes extends to ingredient recovery and upcycling. De Angelis et al. [95] explored the reuse of defective gluten-free pasta made from legumes. They obtained starch-protein-rich fractions with improved water absorption and gel formation properties through dry fractionation. These complexes structurally protected starch, making it less susceptible to gelatinization and thermal damage. This suggests their use in baked goods, sauces, and creams with enhanced functionality and nutritional value.

Another innovative application is in encapsulating bioactive compounds. Qi et al. [96] studied the formation of starch-protein complexes via acid gelation to encapsulate curcumin. By combining starches (common and waxy corn) with isolated proteins (soybean and red bean), they identified hydrophobic interactions and hydrogen bonding that improved the thermal and chemical stability of the gel. During simulated digestion, waxy starches enabled the controlled release of curcumin in the intestine, increasing its bioavailability. This study highlights the potential of these complexes for the efficient delivery of bioactive compounds in functional foods.

On the other hand, protein-starch complexes are being utilized to modify starch digestibility in specific foods. Liu et al. [97] found that adding exogenous proteins, such as wheat protein, egg albumin, and whey protein isolate, to sorghum noodles formed dense networks that hindered enzymatic access. Wheat protein was particularly effective, increasing resistant starch content and slowing hydrolysis, thereby reducing the glycemic index of the noodles.

Emerging technologies, such as 3D printing, have further expanded the possibilities of these complexes. Charvajal-Mena et al. [98] investigated 3D-printed food matrices using salmon-derived proteins. These complexes formed compact structures that increased resistant starch content and reduced digestibility. The molecular interactions enhanced structural stability, making them an innovative tool for designing personalized foods with targeted nutritional benefits.

In a recent study, Cheng et al. [99] investigated the development of protein-starch complexes between soy glycinin (11S) and potato starch using thermal cross-linking. The resulting 11S-potato starch complexes, particularly at a ratio of 2:15, displayed a dense three-dimensional network stabilized by hydrogen bonds and hydrophobic interactions, as confirmed by FTIR and SEM analyses. These complexes were applied as edible coatings on truss tomatoes, significantly improving their barrier properties against moisture and oxygen. Additionally, the coatings reduced weight loss, preserved ascorbic acid and total phenolic content, and extended the shelf life of tomatoes from 7 to 14 days. This research underscores the potential of protein-starch complexes as effective systems for extending the freshness of perishable products.

4.6 Challenges and Opportunities

Stabilizing proteins with starch offers intriguing challenges and opportunities for enhancing food and pharmaceutical applications. Proteins, known for their complex structures and susceptibility to environmental changes, often require stabilization to maintain functionality under various conditions [7, 14]. Starch, a naturally abundant and versatile biopolymer, presents unique capabilities for this purpose, primarily due to its ability to form protective matrices that reduce protein denaturation and aggregation [19, 100, 101]. However, understanding and maximizing the potential of starch-protein complexes involve addressing specific technical and practical challenges.

One significant challenge is the inherent complexity of interactions between proteins and starch molecules, which are influenced by pH, temperature, and ionic strength [8, 37, 38]. Being sensitive to environmental changes, proteins can experience structural destabilization that impacts functionality [62]. However, starch varies in its molecular composition and properties depending on its botanical origin, processing, and preparation methods [102, 103]. This variability can influence the interaction with proteins, making it challenging to design universally stable starch-protein complexes that work across different applications [65, 104, 105]. For example, high temperatures can cause starch to gelatinize, which may benefit or hinder protein encapsulation depending on the application and desired stability parameters [106].

Moreover, the encapsulation and stabilization of proteins in starch-based systems are still underexplored in large-scale production, which poses a challenge for industrial applications. Effective encapsulation techniques are essential to prevent protein aggregation, particularly under high temperatures and pH fluctuations [54, 107]. Recent research highlights the potential of modified starches, such as octenyl succinic anhydride-modified starch, which has improved emulsification properties and enhanced protein stability under thermal processing [108, 109]. However, optimizing such modifications is crucial, as different starch types and protein sources respond differently to chemical treatments. Ensuring these modifications meet food safety and regulatory standards further complicates the process, requiring continuous innovation in green chemistry and regulatory alignment.

Another opportunity lies in developing advanced analytical techniques to characterize starch-protein interactions at the molecular level. Techniques like Fourier-transform infrared spectroscopy (FTIR), nuclear magnetic resonance (NMR), and small-angle X-ray scattering (SAXS) allow researchers to study the bonding types, interaction forces, and structural changes in these complexes [110, 111]. Such analytical advancements can provide insights into optimizing starch modifications, improving protein encapsulation efficiency, and enhancing stability.

In addition to these technical considerations, there is an increasing emphasis on using natural and sustainable stabilizers [112]. Starch, being biodegradable and widely available, aligns well with the growing demand for eco-friendly food ingredients [113]. This opportunity also introduces a challenge: starches must be optimized not only for performance but also for minimal environmental impact, ensuring that the sourcing, modification, and application processes contribute to a sustainable production cycle. Recent research on using deep eutectic solvents for protein extraction illustrates an innovative, environmentally friendly approach to preparing starch-protein complexes [114, 115]. These solvents reduce the need for traditional solvents and can help retain protein functionality, but further optimization is required to address scalability and cost concerns.

Finally, starch's versatility opens the door to potential applications beyond food, including pharmaceuticals and biomedicine, where protein stabilization is essential for therapeutic efficacy [116, 117]. Starch-protein complexes can offer controlled-release functionalities, which are critical in drug delivery systems, by protecting active proteins from degradation until they reach target sites in the body [114, 115]. However, exploring these applications demands rigorous testing and validation to ensure biocompatibility and safety, which can be costly and time-consuming.

While stabilizing proteins with starch presents exciting possibilities, optimizing these systems for various applications requires a deep understanding of starch-protein interactions, advances in green processing methods, and a focus on sustainable production practices. Addressing these challenges can pave the way for starch-based stabilizers to become a staple in food and pharmaceutical industries, fulfilling the demand for natural, efficient, and environmentally responsible solutions.

4.7 Conclusions

Starch-protein interactions present a promising solution to enhance protein stability and functionality under adverse conditions, making them suitable for applications in food and pharmaceutical industries. Starch's chemical and physical modifications expand its utility, particularly in forming protective matrices for functional foods and drug delivery systems. However, challenges such as starch variability and process optimization must be addressed to maximize the efficiency of these complexes. Advances in sustainable starch modification methods and molecular interaction analysis will drive innovation in this field, enabling broader commercial applications that align with sustainability goals and industry demands.

Acknowledgments W.D.C. Chacon gratefully acknowledges the Fundação de Amparo à Pesquisa e Inovação do Estado de Santa Catarina (FAPESC) for the PhD fellowship (3003/2021). G.D. Sorita acknowledges the support of the National Council for Scientific and Technological Development (CNPq, 12/2020 MAI/DAI).

Conflicts of Interest The authors declare no conflict of interest.

References

1. Škrbić T, Giacometti A, Hoang TX, Maritan A, Banavar JR. Amino-acid characteristics in protein native state structures. Biomolecules. 2024;14(7):805.
2. Chen X, Luo J, Fu L, Cai D, Lu X, Liang Z, et al. Structural, physicochemical, and digestibility properties of starch-soybean peptide complex subjected to heat moisture treatment. Food Chem. 2019;297:124957.
3. Nayik GA, Gull A, Masoodi L, Navaf M, Sunooj KV, Ucak İ, et al. Milk proteins: chemistry, functionality and diverse industrial applications. Cogent Food Agric. 2024;10. Informa Healthcare.
4. Wang W, Roberts CJ. Protein aggregation – mechanisms, detection, and control. Int J Pharm. 2018;550:251–68. Elsevier B.V.
5. Ebrahimi SB, Samanta D. Engineering protein-based therapeutics through structural and chemical design. Nat Commun. 2023;14:2411. Nature Research.
6. Eidenberger L, Kogelmann B, Steinkellner H. Plant-based biopharmaceutical engineering. Nat Rev Bioeng. 2023;1(6):426–39.
7. Acharya VV, Chaudhuri P. Modalities of protein denaturation and nature of denaturants. Int J Pharm Sci Rev Res. 2021;69(2):19–24.
8. Ballauff M. Denaturation of proteins: electrostatic effects vs. hydration. RSC Adv. 2022;12(16):10105–13.
9. Anema SG. Heat-induced changes in caseins and casein micelles, including interactions with denatured whey proteins. Int Dairy J. 2021;122:105136. Elsevier Ltd.
10. Housmans JAJ, Wu G, Schymkowitz J, Rousseau F. A guide to studying protein aggregation. FEBS J. 2023;290:554–83. John Wiley and Sons Inc.
11. Xu Y, Sun L, Zhuang Y, Gu Y, Cheng G, Fan X, et al. Protein-stabilized emulsion gels with improved emulsifying and gelling properties for the delivery of bioactive ingredients: a review. Foods. 2023;12:2703. Multidisciplinary Digital Publishing Institute (MDPI).

12. Li M, Ritzoulis C, Du Q, Liu Y, Ding Y, Liu W, et al. Recent progress on protein-polyphenol complexes: effect on stability and nutrients delivery of oil-in-water emulsion system. Front Nutr. 2021;8:765589. Frontiers Media S.A.
13. Rahban M, Ahmad F, Piatyszek MA, Haertlé T, Saso L, Saboury AA. Stabilization challenges and aggregation in protein-based therapeutics in the pharmaceutical industry. RSC Adv. 2023;13:35947–63. Royal Society of Chemistry.
14. Akbarian M, Chen SH. Instability challenges and stabilization strategies of pharmaceutical proteins. Pharmaceutics. 2022;14:2533. MDPI.
15. Sun X, Wang H, Li S, Song C, Zhang S, Ren J, et al. Maillard-type protein–polysaccharide conjugates and electrostatic protein–polysaccharide complexes as delivery vehicles for food bioactive ingredients: formation, types, and applications. Gels. 2022;8:135. MDPI.
16. Xiao Y, Ahmad T, Belwal T, Aadil RM, Siddique M, Pang L, et al. A review on protein based nanocarriers for polyphenols: interaction and stabilization mechanisms. Food Innov Adv. 2023;2(3):193–202.
17. Nooshkam M, Varidi M, Zareie Z, Alkobeisi F. Behavior of protein-polysaccharide conjugate-stabilized food emulsions under various destabilization conditions. Food Chem X. 2023;18:100725. Elsevier Ltd.
18. Tong L, Wang L. Production and properties of starch: current research. Molecules. 2024;29(3):646.
19. Apostolidis E, Stoforos GN, Mandala I. Starch physical treatment, emulsion formation, stability, and their applications. Carbohydr Polym. 2023;305:120554. Elsevier Ltd.
20. Mahfouzi M, Zhang H, Haoran L, McClements DJ, Hadidi M. Starch-based particles as stabilizers for Pickering emulsions: modification, characteristics, stabilization, and applications. Crit Rev Food Sci Nutr. Taylor and Francis Ltd. 2025;65(10):1841–56.
21. Yu B, Ren F, Zhao H, Cui B, Liu P. Effects of native starch and modified starches on the textural, rheological and microstructural characteristics of soybean protein gel. Int J Biol Macromol. 2020;142:237–43.
22. Zhao C, Miao Z, Qi Q, Zheng Q, Mao Y, Chu Z, et al. Interactions of soy protein isolate with common and waxy corn starches and their effects on acid-induced cold gelation properties of complexes. Food Chem X. 2023;18:100671.
23. Chen D, Fang F, Federici E, Campanella O, Jones OG. Rheology, microstructure and phase behavior of potato starch-protein fibril mixed gel. Carbohydr Polym. 2020;239:116247.
24. Zhu Y, Tao H, Janaswamy S, Zou F, Cui B, Guo L. The functionality of laccase- or peroxidase-treated potato flour: role of interactions between protein and protein/starch. Food Chem. 2021;341:128082.
25. Lemos PVF, Marcelino HR, Cardoso LG, de Souza CO, Druzian JI. Starch chemical modifications applied to drug delivery systems: from fundamentals to FDA-approved raw materials. Int J Biol Macromol. Elsevier B.V. 2021;184:218–34.
26. Otache MA, Duru RU, Achugasim O, Abayeh OJ. Advances in the modification of starch via esterification for enhanced properties. J Polym Environ. Springer. 2021;29:1365–79.
27. Morris R, Black KA, Stollar EJ. Uncovering protein function: from classification to complexes. Essays Biochem. 2022;66:255–85. Portland Press Ltd.
28. Cipak L. Protein kinases: function, substrates, and implication in diseases. Int J Mol Sci. 2022;23:3560. MDPI.
29. Popescu F, Titorencu I, Albu Kaya M, Miculescu F, Tutuianu R, Coman AE, et al. Development of innovative biocomposites with collagen, keratin and hydroxyapatite for bone tissue engineering. Biomimetics (Basel). 2024;9(7):428.
30. Wuyun Q, Chen Y, Shen Y, Cao Y, Hu G, Cui W, et al. Recent progress of protein tertiary structure prediction. Molecules. 2024;29:832. Multidisciplinary Digital Publishing Institute (MDPI).
31. Ma Z, Chen J, Wang C. Research on protein tertiary structure generation [Internet]. Int J Eng Inventions. 2023;12. Available from: www.ijeijournal.com.

32. Marciano S, Dey D, Listov D, Fleishman SJ, Sonn-Segev A, Mertens H, et al. Protein quaternary structures in solution are a mixture of multiple forms. Chem Sci. 2022;13(39):11680–95.
33. Ramanathan A, Agarwal PK. Evolutionarily conserved linkage between enzyme fold, flexibility, and catalysis. PLoS Biol. 2011;9(11):e1001193.
34. Ding Y, Zhang S, Hess H, Kong X, Zhang Y. Replicating enzymatic activity by positioning active sites with synthetic protein scaffolds [Internet]. 2024. Available from: http://biorxiv.org/lookup/doi/10.1101/2024.01.31.577620.
35. Cao J, Tong X, Cao X, Peng Z, Zheng L, Dai J, et al. Effect of pH on the soybean whey protein–gum arabic emulsion delivery systems for curcumin: emulsifying, stability, and digestive properties. Food Chem. 2024;456:139938.
36. Hazt B, Bassani HP, Elias-Machado JP, Aldinucci Buzzo JL, Silveira JLM, de Freitas RA. Effect of pH and protein particle shape on the stability of amylopectin–xyloglucan water-in-water emulsions. Food Hydrocoll. 2020;104:105769.
37. Timr S, Madern D, Sterpone F. Protein thermal stability. In: Progress in molecular biology and translational science. Amsterdam: Elsevier B.V; 2020. p. 239–72.
38. Hamborg L, Horsted EW, Johansson KE, Willemoës M, Lindorff-Larsen K, Teilum K. Global analysis of protein stability by temperature and chemical denaturation. Anal Biochem. 2020;605:113863.
39. Infante E, Stannard A, Board SJ, Rico-Lastres P, Rostkova E, Beedle AEM, et al. The mechanical stability of proteins regulates their translocation rate into the cell nucleus. Nat Phys. 2019;15(9):973–81.
40. Koyama S, Nemoto Y, Ichikawa M, Oka D, Tsujii Y, Noguchi T, et al. Effects of suppressing protein structural changes on the excellent gelling properties of dried egg white via dry-heat treatment. Food Sci Technol Res. 2021;27(2):293–300.
41. Zuo J, Zheng W, Shi N, Song R, Han F, Yang C, et al. Study on the thermal stability of the sweet-tasting protein Brazzein based on its structure-sweetness relationship. J Agric Food Chem. 2024;72(13):7374–82.
42. Singh J, Dean A, Prakash S, Bhandari B, Bansal N. Ultra high temperature stability of milk protein concentrate: effect of mineral salts addition. J Food Eng. 2021;300:110503.
43. Kim HJ, Lee JH, Hur YB, Lee CW, Park SH, Koo BW. Marine antifreeze proteins: structure, function, and application to cryopreservation as a potential cryoprotectant. Mar Drugs. 2017;15(2):27.
44. Rosa M, Roberts CJ, Rodrigues MA. Connecting high-temperature and low-temperature protein stability and aggregation. PLoS One. 2017;12(5):e0176748.
45. Holeček M. Aspartic acid in health and disease. Nutrients. 2023;15(18):4023. Multidisciplinary Digital Publishing Institute (MDPI).
46. Dharmayanti C, Gillam TA, Klingler-Hoffmann M, Albrecht H, Blencowe A. Strategies for the development of pH-responsive synthetic polypeptides and polymer-peptide hybrids: recent advancements. Polymers (Basel). 2021;13(4):1–17.
47. Hung CL, Kuo YH, Lee SW, Chiang YW. Protein stability depends critically on the surface hydrogen-bonding network: a case study of bid protein. J Phys Chem B. 2021;125(30):8373–82.
48. Magsumov T, Ziying L, Sedov I. Comparative study of the protein denaturing ability of different organic cosolvents. Int J Biol Macromol. 2020;160:880–8.
49. Wang Z, Sun F. Effects of processing technology on protein separation, quality, and functional characteristics. Foods. 2023;12:3841. Multidisciplinary Digital Publishing Institute (MDPI).
50. Moino C, Artusio F, Pisano R. Shear stress as a driver of degradation for protein-based therapeutics: more accomplice than culprit. Int J Pharm. 2024;650:123679. Elsevier B.V.
51. Gikanga B, Maa YF. A review on mixing-induced protein particle formation: the puzzle of bottom-mounted mixers. J Pharm Sci. 2020;109:2363–74. Elsevier B.V.
52. Yang J, Liu G, Zeng H, Chen L. Effects of high pressure homogenization on faba bean protein aggregation in relation to solubility and interfacial properties. Food Hydrocoll. 2018;83:275–86.

53. Anema SG. The whey proteins in milk: thermal denaturation, physical interactions, and effects on the functional properties of milk. In: Milk proteins: from expression to food. Amsterdam: Elsevier; 2019. p. 325–84.
54. Rajan R, Ahmed S, Sharma N, Kumar N, Debas A, Matsumura K. Review of the current state of protein aggregation inhibition from a materials chemistry perspective: special focus on polymeric materials. Mater Adv. 2021;2:1139–76. Royal Society of Chemistry.
55. Dos Santos SL, Petropoulos I, Friguet B. The oxidized protein repair enzymes methionine sulfoxide reductases and their roles in protecting against oxidative stress, in ageing and in regulating protein function. Antioxidants (Basel). 2018;7:191. MDPI
56. Liang X, Kaya A, Zhang Y, Le DT, Hua D, Gladyshev VN. Characterization of methionine oxidation and methionine sulfoxide reduction using methionine-rich cysteine-free proteins. BMC Biochem. 2012;13(1):21.
57. Percio A, Cicchinelli M, Masci D, Summo M, Urbani A, Greco V. Oxidative cysteine post translational modifications drive the redox code underlying neurodegeneration and amyotrophic lateral sclerosis. Antioxidants (Basel). 2024;13:883. Multidisciplinary Digital Publishing Institute (MDPI).
58. Müller-Santos M, Koskimäki JJ, Alves LPS, De Souza EM, Jendrossek D, Pirttilä AM. The protective role of PHB and its degradation products against stress situations in bacteria. FEMS Microbiol Rev. 2021;45:fuaa058. Oxford University Press.
59. Xiong YL, Guo A. Animal and plant protein oxidation: chemical and functional property significance. Foods. 2021;10:40. MDPI AG.
60. Rong L, Shen M, Zhang Y, Yu H, Xie J. Food polysaccharides and proteins: processing, characterization, and health benefits. Foods. 2024;13(7):1113. Multidisciplinary Digital Publishing Institute (MDPI).
61. Louros N, Schymkowitz J, Rousseau F. Mechanisms and pathology of protein misfolding and aggregation. Nat Rev Mol Cell Biol. 2023;24(12):912–33. Nature Research.
62. Wijayanti HB, Brodkorb A, Hogan SA, Murphy EG. Thermal denaturation, aggregation, and methods of prevention. In: Whey proteins: from milk to medicine. Amsterdam: Elsevier; 2018. p. 185–247.
63. Karunnanithy V, Abdul Rahman NHB, Abdullah NAH, Fauzi MB, Lokanathan Y, Min Hwei AN, et al. Effectiveness of lyoprotectants in protein stabilization during lyophilization. Pharmaceutics. 2024;16:1346. Multidisciplinary Digital Publishing Institute (MDPI).
64. Liu FF, Ji L, Zhang L, Dong XY, Sun Y. Molecular basis for polyol-induced protein stability revealed by molecular dynamics simulations. J Chem Phys. 2010;132(22):225103.
65. Zhu P, Du X, Liu C, Zhao G, Wang M. Effects of pH during dry-heat preparation on the physicochemical and emulsifying properties of rice starch and whey protein isolate mixtures. Food Hydrocoll. 2023;140:108614.
66. Zhen Y, Wang K, Wang J, Qiao D, Zhao S, Lin Q, et al. Increasing the pH value during thermal processing suppresses the starch digestion of the resulting starch-protein-lipid complexes. Carbohydr Polym. 2022;278:118931.
67. Zhan Q, Ye X, Zhang Y, Kong X, Bao J, Corke H, et al. Starch granule-associated proteins affect the physicochemical properties of rice starch. Food Hydrocoll. 2020;101:105504.
68. Yu L, Ma Y, Zhao Y, Rehman A-u, Guo L, Liu Y, et al. Interaction of B-type starch with gluten skeleton improves wheat dough mixing properties by stabilizing gluten micro-structure. Food Chem. 2022;371:131390.
69. Scott G, Awika JM. Effect of protein–starch interactions on starch retrogradation and implications for food product quality. Compr Rev Food Sci Food Saf [Internet]. 2023;22(3):2081–111. [cited 2024 Dec 3]. Available from: https://onlinelibrary.wiley.com/doi/full/10.1111/1541-4337.13141
70. Peng P, Wang X, Zou X, Zhang X, Hu X. Dynamic behaviors of protein and starch and interactions associated with glutenin composition in wheat dough matrices during sequential thermo-mechanical treatments. Food Res Int. 2022;154:110986.

71. Bravo-Núñez Á, Garzón R, Rosell CM, Gómez M. Evaluation of starch–protein interactions as a function of pH. Foods. 2019;8:155. [Internet]. 2019 May 7 [cited 2024 Dec 3];8(5):155. Available from: https://www.mdpi.com/2304-8158/8/5/155/htm.
72. Farooq MA, Yu J. Recent advances in physical processing techniques to enhance the resistant starch content in foods: a review. Foods. 2024;13:2770. [Internet]. 2024 Aug 30 [cited 2024 Dec 3];13(17):2770. Available from: https://www.mdpi.com/2304-8158/13/17/2770/htm.
73. Liu S, Chen W, Zhang C, Wu T, Zheng B, Guo Z. Structural, thermal and pasting properties of heat-treated lotus seed starch–protein mixtures. Foods. 2022;11:2933. [Internet]. 2022 Sep 20 [cited 2024 Dec 3];11(19):2933. Available from: https://www.mdpi.com/2304-8158/11/19/2933/htm.
74. Wang J, Zhao S, Min G, Qiao D, Zhang B, Niu M, et al. Starch-protein interplay varies the multi-scale structures of starch undergoing thermal processing. Int J Biol Macromol. 2021;175:179–87.
75. Yang C, Zhong F, Douglas Goff H, Li Y. Study on starch-protein interactions and their effects on physicochemical and digestible properties of the blends. Food Chem. 2019;280:51–8.
76. Wang S, Chao C, Cai J, Niu B, Copeland L, Wang S. Starch–lipid and starch–lipid–protein complexes: a comprehensive review. Compr Rev Food Sci Food Saf [Internet]. 2020;19(3):1056–79. [cited 2024 Dec 3]. Available from: https://drive.google.com/file/d/1Yatqon59S8ljE33fE8NjU-TYR7aqouHx/view?usp=drive_web&usp=embed_facebook.
77. Budai L, Budai M, Fülöpné Pápay ZE, Vilimi Z, Antal I. Rheological considerations of pharmaceutical formulations: focus on viscoelasticity. Gels. 2023;9:469. [Internet]. 2023 Jun 7 [cited 2024 Dec 3];9(6):469. Available from: https://www.mdpi.com/2310-2861/9/6/469/htm.
78. Fitzpatrick AW, Knowles TPJ, Waudby CA, Vendruscolo M, Dobson CM. Inversion of the balance between hydrophobic and hydrogen bonding interactions in protein folding and aggregation. PLoS Comput Biol [Internet]. 2011;7(10):e1002169. [cited 2024 Dec 3]. Available from: https://pubmed.ncbi.nlm.nih.gov/22022239/.
79. Pei Y, Deng Z, Li B. Regulation on aggregation behavior and in vitro digestibility of phytic acid–whey protein isolate complexes: effects of heating, pH and phytic acid levels. Foods. 2024;13:3491. [Internet]. 2024 [cited 2024 Dec 3];13(21):3491. Available from: https://www.mdpi.com/2304-8158/13/21/3491/htm
80. Adal E, Aktar T, Keskin Çavdar H. Alteration of the morphological and physicochemical characteristics of corn and wheat starch via dry heating with whey protein isolates. Foods. 2024;13:3701. [Internet]. 2024 Nov 20 [cited 2024 Dec 3];13(22):3701. Available from: https://www.mdpi.com/2304-8158/13/22/3701/htm
81. Woo SH, Kim JS, Jeong HM, Shin YJ, Hong JS, Choi HD, et al. Development of freeze-thaw stable starch through enzymatic modification. Foods. 2021;10:2269. [Internet]. 2021 Sep 25 [cited 2024 Dec 3];10(10):2269. Available from: https://www.mdpi.com/2304-8158/10/10/2269/htm
82. Lu X, Zhan J, Ma R, Tian Y. Structure, thermal stability, and in vitro digestibility of rice starch–protein hydrolysate complexes prepared using different hydrothermal treatments. Int J Biol Macromol. 2023;230:123130.
83. XIE S h, WANG Z j, HE Z y, ZENG M m, QIN F, ADHIKARI B, et al. The effects of maltodextrin/starch in soy protein isolate–wheat gluten on the thermal stability of high-moisture extrudates. J Integr Agric. 2023;22(5):1590–602.
84. Allen KE, Carpenter CE, Walsh MK. Influence of protein level and starch type on an extrusion-expanded whey product. Int J Food Sci Technol [Internet]. 2007;42(8):953–60. [cited 2024 Dec 3]. Available from: https://onlinelibrary.wiley.com/doi/full/10.1111/j.1365-2621.2006.01316.x.
85. Chu L, Yang L, Li J, Lin L, Zheng G. Effect of Smilax china L. starch on the gel properties and interactions of calcium sulfate-induced soy protein isolate gel. Int J Biol Macromol. 2019;135:127–32.

86. Du J, Lv M y, Zhang H l, Xiao S s, Zheng S y, Wang X d. Synergistic effect of endogenous gluten and oleic acid on wheat starch digestion by forming ordered starch-fatty acid-protein complexes during thermal processing. Curr Res Food Sci. 2023;6:100422.
87. Dong D, Cui B. Fabrication, characterization and emulsifying properties of potato starch/soy protein complexes in acidic conditions. Food Hydrocoll. 2021;115:106600.
88. Lin L, Yu X, Gao Y, Mei L, Zhu Z, Du X. Physicochemical properties and in vitro starch digestibility of wheat starch/rice protein hydrolysate complexes. Food Hydrocoll. 2022;125:107348.
89. Lu X, Ma R, Qiu H, Sun C, Tian Y. Mechanism of effect of endogenous/exogenous rice protein and its hydrolysates on rice starch digestibility. Int J Biol Macromol. 2021;193:311–8.
90. Scott G, Awika JM. Effect of protein–starch interactions on starch retrogradation and implications for food product quality. Compr Rev Food Sci Food Saf. 2023;22:2081–111. John Wiley and Sons Inc
91. Dobson S, Laredo T, Marangoni AG. Particle filled protein-starch composites as the basis for plant-based meat analogues. Curr Res Food Sci. 2022;5:892–903.
92. Shi M, Dong X, Cheng Y, Ji X, Liu Y, Yan Y. Preparation and characterization of extruded yam starch–soy protein isolate complexes and their effects on the quality of dough. Foods. 2023;12(2):360.
93. Zeng F, Hu Z, Yang Y, Jin Z, Jiao A. Baking quality and starch digestibility in whole wheat bread based on protein enrichment strategy: significance of protein-starch-water interaction in dough. Food Biosci. 2024;61:104403.
94. Liu S, Zhao L, Zhang J, Wang L, Liu H. Functional drink powders from vertical-stone-milled oat and highland barley with high dietary-fiber levels decrease the postprandial glycemic response. J Funct Foods. 2021;83:104548.
95. De Angelis D, Pasqualone A, Manfredi L, Allegretta I, Terzano R, Summo C. Dry fractionation as a promising technology to reuse the physically defected legume-based gluten-free pasta. Int J Food Sci Technol. 2022;57(8):4816–24.
96. Qi Q, Yan X, Mao Y, Wang F, Zhang H, Xu X, et al. Fabrication and characterization of curcumin-loaded complex acid-induced cold gels of soybean and red bean protein isolates incorporated with common and waxy corn starches. Food Hydrocoll. 2024;151:109863.
97. Liu FY, Yang Z, Guo XN, Xing JJ, Zhu KX. Influence of protein type, content and polymerization on in vitro starch digestibility of sorghum noodles. Food Res Int. 2021;142:110199.
98. Carvajal-Mena N, Tabilo-Munizaga G, Pérez-Won M, Herrera-Lavados C, Moreno-Osorio L. Influence of starch-protein interactions on the digestibility and chemical properties of a 3D-printed food matrix based on salmon by-product proteins. Food Res Int. 2024;179:114035.
99. Cheng X, Yang S, Fang Q, Dai S, Peng X, Sun M, et al. Biomacromolecule assembly of soy glycinin-potato starch complexes: focus on structure, function, and applications. Carbohydr Polym. 2023;317:121101.
100. Bravo-Núñez Á, Garzón R, Rosell CM, Gómez M. Evaluation of starch-protein interactions as a function of pH. Foods. 2019;8(5):155.
101. Cai J, Chao C, Niu B, Copeland L, Yu J, Wang S, et al. New insight into the interactions among starch, lipid and protein in model systems with different starches. Food Hydrocoll. 2021;112:106323.
102. Dereje B. Composition, morphology and physicochemical properties of starches derived from indigenous Ethiopian tuber crops: a review. Int J Biol Macromol. 2021;187:911–21. Elsevier B.V.
103. Apriyanto A, Compart J, Fettke J. A review of starch, a unique biopolymer – structure, metabolism and in planta modifications. Plant Sci. 2022;318:111223. Elsevier Ireland Ltd.
104. Zheng D, Huang C, Li B, Zhu X, Liu R, Zhao H. Effect of a novel shell material—starch-protein-fatty acid ternary nanoparticles on loading levels and in vitro release of curcumin. Int J Biol Macromol. 2021;192:471–8.

105. Zhang W, Zhao D, Dong Z, Li J, Zhang B, Yu W. The consistency factor and the viscosity exponent of soybean-protein-isolate/wheat-gluten/corn-starch blends by using a capillary rheometry. Molecules. 2022;27(19):6693.
106. Suri S, Singh A. Modification of starch by novel and traditional ways: influence on the structure and functional properties. Sustain Food Technol. 2023;1:348–62. Royal Society of Chemistry.
107. Zhao M, Zacharia NS. Protein encapsulation via polyelectrolyte complex coacervation: protection against protein denaturation. J Chem Phys. 2018;149(16):163326.
108. Sivabalan S, Ross CF, Tang J, Sablani SS. Physical, thermal, and storage stability of multilayered emulsion loaded with β-carotene. Food Innov Adv. 2024;3(3):244–55.
109. Zhou J, Guo M, Qin Y, Wang W, Lv R, Xu E, et al. Advances in starch nanoparticle for emulsion stabilization. Foods. 2023;12:2425. MDPI.
110. Tatulian SA. Analysis of protein-protein and protein-membrane interactions by isotope-edited infrared spectroscopy. Phys Chem Chem Phys. 2024;26:21930–53. Royal Society of Chemistry.
111. Sun Y, Li X, Chen R, Liu F, Wei S. Recent advances in structural characterization of biomacromolecules in foods via small-angle X-ray scattering. Front Nutr. 2022;9:1039762. Frontiers Media S.A.
112. Martau GA, Mihai M, Vodnar DC. The use of chitosan, alginate, and pectin in the biomedical and food sector-biocompatibility, bioadhesiveness, and biodegradability. Polymers (Basel). 2019;11:1837. MDPI AG.
113. Gałkowska D, Kapuśniak K, Juszczak L. Chemically modified starches as food additives. Molecules. 2023;28:7543. Multidisciplinary Digital Publishing Institute (MDPI).
114. Bowen H, Durrani R, Delavault A, Durand E, Chenyu J, Yiyang L, et al. Application of deep eutectic solvents in protein extraction and purification. Front Chem. 2022;10:912411. Frontiers Media S.A.
115. Kumorow AC, Wardhani DH, Kusworo TD, Djaeni M, Ping TC, Alhanif M. Deep eutectic solvents and natural deep eutectic solvents for extraction and purification of proteins from animal and botanical sources - a review. Int J Technol. 2024;15:1420–37. Faculty of Engineering, Universitas Indonesia.
116. Wiącek AE, Furmaniuk A. Starch-based polysaccharide systems with bioactive substances: physicochemical and wettability characteristics. Int J Mol Sci. 2024;25(9):4590.
117. Lee CS, Hwang HS. Starch-based hydrogels as a drug delivery system in biomedical applications. Gels. 2023;9:951. Multidisciplinary Digital Publishing Institute (MDPI).

Chapter 5
Starch-Based Food Ingredients with Low Glycemic Index

Perla Osorio-Diaz, Edith Agama-Acevedo, Maria A. Hernandez-Aguirre, and Luis A. Bello-Perez

5.1 Introduction

Consumers demand healthy foods with low glycemic index (GI), such as bakery products, pasta, snacks, and cookies, among others. However, traditional foods in those groups are nutritionally poor. After all, they use refined cereals and pseudocereal grain flours and starches in their formulations [1] because their consumption results in high glucose levels in the blood. These foods have been related to the prevalence of health problems, such as overweight, obesity, diabetes, and cardiovascular diseases. The starch hydrolysis rate and absorption in the small intestine influence the postprandial glycemic response, which, over time, is associated with blood glucose abnormalities, insulin response, appetitive response, and food intake [2].

The development of functional foods with low GI (low carb) and low fat is an opportunity area in the food industry. Diverse strategies have been proposed to regulate glucose homeostasis, such as a reduction in raw materials with highly digestible starch (complete gelatinization), an increase in dietary fiber content (whole grains), and processing that completely disorganizes the starch structure (highly prone to hydrolysis by digestive enzymes).

The healthy foods market in the world has shown a growing trend in the last few years. The North American market size is estimated at 11.76B in 2024 and is expected to grow to around 15.38B by 2030 [3]. More information in other markets is shown in Table 5.1. Traditional and healthy foods are elaborated with diverse ingredients, the main one being starchy flour which can be gluten-free (Table 5.2). Starch is the most important macromolecule in foods and normally requires a thermal process (cooking, baking, steaming, extrusion, or frying) before inclusion in the human diet. The intensity in the thermal process produces different levels of starch

P. Osorio-Diaz (✉) · E. Agama-Acevedo · M. A. Hernandez-Aguirre · L. A. Bello-Perez
Instituto Politécnico Nacional, CEPROBI, Yautepec, Mexico
e-mail: posorio@ipn.mx

G. Ayala Valencia (ed.), *Starch: Progress in Food Applications*,
https://doi.org/10.1007/978-3-031-98732-8_5

Table 5.1 The market for healthy starchy foodstuffs

Region	Country	Group	Product	Source	Profits (year)	Calculated growing	Profits in 2030	References
North America	U.S., Canada, Mexico	Healthy Snack Chips	Tortilla, pop, popcorn, puff, potato, and tapioca chips, and others	Maize Potato Tapioca Others	(2024) USD 11.76 billion	CAGR of 4.6% from 2024 to 2030	USD 15.38 billion	[4]
Global Latin America, Europe, North America, Middle East & Africa, Asia Pacific	Canada Mexico U.S. France Germany Italy Spain UK Australia China India Japan Brazil Argentina Saudi Arabia	Bakery & confectionery		Corn Wheat Cassava Potato Others	(US$M) 24,278.3	Estimated to grow at a compound annual growth rate (CAGR) of 9% from 2024 to 2030	(US$M) 49,924.9	[5, 6]

Table 5.2 Alternative sources (starch and flour) used as ingredients in foodstuffs

Raw material	Ingredient	Foodstuff	Region	Country	Profits 2023	Calculated growing	Profits 2030	References
Lentils, peas, chickpeas, and others	Pulses (Pulse Flour Pulse Starch Pulse Protein Pulse Fiber & Grits)	Bakery & confectioner, snacks, dairy, Beverages, feed, others	North America; Europe; Asia Pacific; Central and South America; Middle East and Africa	U.S.; Canaca; Mexico; UK; Germany; France; Spain; Italy; China; Japan; South Korea; Australia & New Zealand; India; Brazil; South Africa	USD 22.78 billion (2024)	CAGR of 3.1% from 2024 to 2030	USD 27.30 billion	[7]
Fruits and nuts: Banana, cashew, nut Grains: Corn, wheat, and rice Vegetables: Peas, potato and cassava Cereal Food	Resistant Starch RS1 RS2 RS3 RS4	Bakery Products Bread Biscuits and Cookies Crackers Other Bakery Products Confectionery Beverages Breakfast Cereals Pasta and Noodles Dairy Products Nutrition Bars Meat and Processed Food Meat Substitutes Soups, Dressings, and Condiments Others	Global North America; Europe; Asia Pacific; Central and South America; Middle East & Africa	U.S.; Canada; Mexico; U.K.; Germany; France; Italy; Spain; China; India; Japan; Australia & New Zealand; South Korea; Brazil; Argentina; UAE; South Africa	USD 1370.28 billion (2023)	CAGR of 8.9% from 2023 to 2030	USD 2511.07 billion	[8]

disorganization (gelatinization), which is key to hydrolysis by digestive enzymes. Starch digestion is controlled by its microstructure, which remains after processing, and its rearrangement during storage.

In addition, the components of the food matrix are important to regulate the starch hydrolysis rate because some macromolecules (lipids and proteins), non-starch polysaccharides (hydrocolloids, cellulose, and arabinoxylans), and molecules (polyphenols) can produce interactions with starch to produce complexes. The interactions between starch and macromolecules produce V-type complexes. In lipids, fatty acids are embedded in a hydrophobic cavity [9] for different protein interactions with starch. Starch–zein microencapsulation has been studied to mimic starch–protein matrices that decrease the starch hydrolysis rate [10]. Interactions of starch conjugated with amino acids through the Maillard reaction have been shown to decrease starch hydrolysis [11]. Non-starch polysaccharides and proteins (gluten) produce a network that restricts the access of digestive enzymes to starch for hydrolysis, slowing the hydrolysis rate with a low GI [12]. Cellulose is reported to inhibit amylases and reduce starch hydrolysis rate [13]. The interactions between polyphenols and starch are generated by the inclusion complex in the hydrophobic cavity produced by linear chains of starch where the polyphenol is embedded. The interaction between starch and small polyphenol molecules (e.g., gallic acid) results from a hydrophobic bond, while others, such as hydrogen bonds, electrostatic interaction, and van der Waals forces, are not the main interaction forces producing the complex [14].

Diverse ingredients are essential to producing food matrices and modulating the extent of the starch hydrolysis rate. Traditionally, whole-grain flour has been used to produce foods with low GI. The use of starchy flours with high amylose content (wheat, maize, potato) increases indigestible starch (resistant starch) content given that restriction to swelling is present during cooking, leading to a partial disorganization of starch granules (intact granules), not processed by digestive enzymes. Starch-lipids and starch-polyphenols complexes prepared as functional ingredients can be included in a food matrix formulation or added to ready-to-eat foods. This reduces the amount of available starch and increases the resistant starch (RS) that continues the journey to the colon.

5.2 Starch Digestion Rate

Starch is the main macromolecule that provides biological fuel (45–65% of caloric intake) for the population worldwide [2]. Starch hydrolysis rate in foods has been a topic of interest for over 30 years. It has been accepted that not all starch present in foods is hydrolyzed during the digestion process that begins in the mouth and continues in the small intestine [15]. Starch in foods can be classified as glycemic or nonglycemic because of its bioavailability (Fig. 5.1). The measurement of the postprandial glycemic response is used to determine the GI. Starch digestibility has been associated with GI to determine the nutritional quality of starchy foods. These

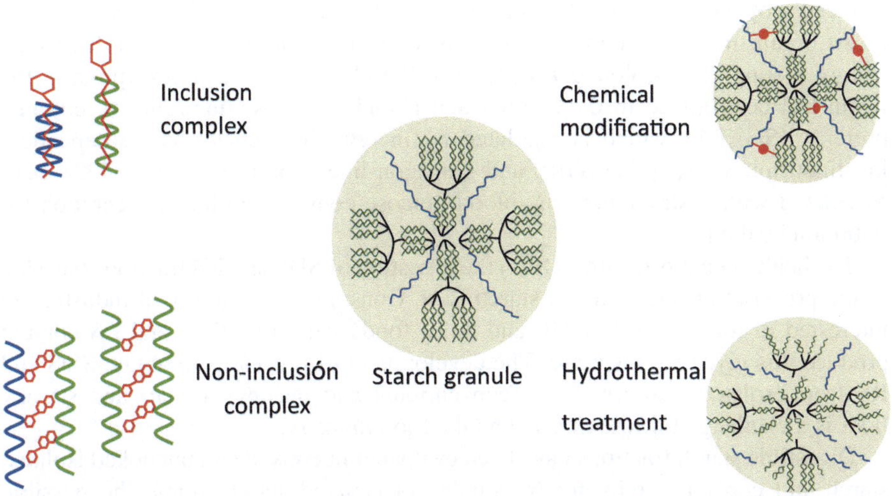

Fig. 5.1 Structure of modified starch by thermal and chemical treatments for complex formation used for elaboration of functional ingredients with low glycemic index

are classified based on the glycemic response [16, 17]. The method to test GI uses subjects supplied with food, blood samples are obtained from different periods, and glucose is quantified. This method is difficult and expensive. There are reports on a good correlation between *in vivo* and *in vitro* methods to estimate the GI of starchy foods [18]. They suggest the *in vitro* method since no subjects are required, making the experimental procedure affordable.

5.2.1 Starch Digestion Fractions

Based on the rate and extent of glucose release, Englyst et al. [19] proposed three fractions: rapidly digestible starch (RDS) fraction, digested within 20 min after ingesting the starchy meal, slowly digestible starch (SDS) digested between 20 and 120 min, and RS fraction, undigested after 120 min. This classification is based on the physiological conditions of the human being because the insulinemic response regulates glucose levels in the blood of healthy people 120 min after digestion. The three fractions have different physiological effects; RDS is rapidly digested in the mouth and the first section of the small intestine (duodenum), producing postprandial hyperglycemia. Slowly digested throughout the entire small intestine, SDS supplies a sustained glucose release along the approximately 30 ft. of the small intestine because no glucose peaks are present and there is a low insulinemic response. It provides a sustained glucose release because the human body has the fuel to carry out physiological functions for long periods. Undigested RS reaches the colon as a substrate for the microbiota and is included as a part of the dietary fiber. It is fermented by the microbiota to produce short-chain fatty acids (SCFAs), beneficial to the organism. For example, butyrate is associated with colon health due to the energy it supplies to colonocytes [20, 21]. Starch digestion and absorption in the ileum has been linked to the activity and abundance of sodium/glucose cotransporter 1 (SGLT-1), which is regulated by incretin hormones glucose-dependent insulinotropic polypeptide (GIP) and glucagon-like peptide-1 (GLP-1) [22], both associated with gastric emptying, blood glucose control, and insulin secretion for optimum health [23].

The health benefits resulting from the ingestion of SDS and RS fractions have led to the proposal of nutraceutical starch [24]. Consumers and the food industry are interested in foods with low GI, and those foods with high SDS and RS content produce this glycemic response. The development of processes to obtain SDS and RS from isolated starch sources (conventional and unconventional) and starchy flour is a challenge for researchers and the food industry.

Digestible starch fractions have been evaluated in cooked and uncooked isolated starch and cooked starchy foods. Studies in isolated starch mimic the possible impact of cooking in food matrices prepared with the raw material [25, 26]. Alternative standard analyses to classify the fractions (RDS, SDS, and RS) based on the hydrolysis rate have been proposed. An improved first-order kinetic model by Goñi et al. [18] uses a logarithm of the slope (LOS) plot [27]. This approach allows an estimation of several digestion kinetic parameters: k (constant rate) is represented by the negative slope of the linear plot, and the digested total starch is obtained from the intercept of the y-axis. The analysis showed that the digestion of raw native starch does not follow a single first-order reaction; the digestion process is described by two separate first-order reactions with different k values. The LOS plot approach would be useful for the accurate determination of RDS and SDS fractions. An extension of this model considered two starch fractions, one linked to a fast digestion rate and the other to a slow digestion. The experimental data were fitted by numerically solving a non-linear least-squares problem. The estimated parameters were interpreted in terms of reaction rates and digestible/RS fractions [28]. The laboratory analysis and the fit of the experimental data using the models discussed here are useful to obtain information on starch digestion in food matrices.

5.3 Starch-Based Ingredients

Starch-based ingredients are widely used in the production of foodstuffs due to their functional and sensorial characteristics; however, their consumption should be moderate due to the highly caloric content they supply, which is associated with obesity, diabetes, cardiovascular problems, and metabolic syndrome. Several starchy-based ingredients are considered healthy since they are not digested in the small intestine and release some compounds (e.g. antioxidants) that provide some health benefits. Figure 5.2 shows the starch-based ingredients discussed in this chapter.

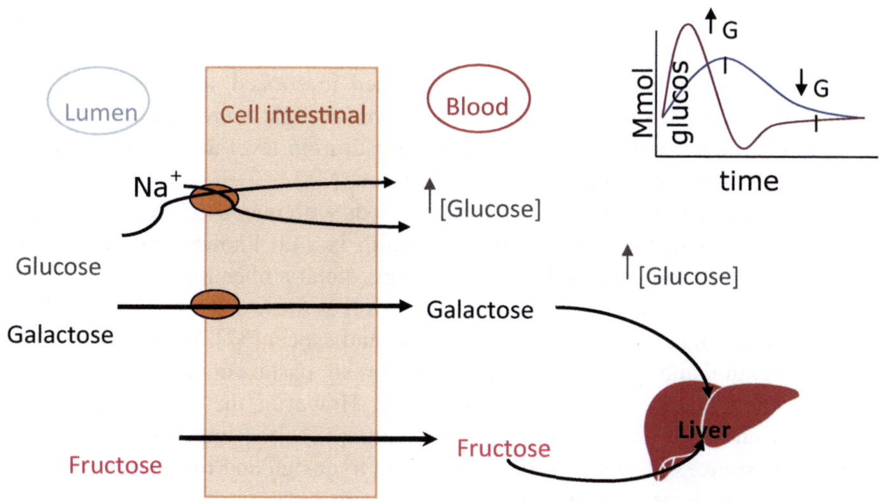

Fig. 5.2 Effect of starch hydrolysis on the glucose absorption in the small intestine and the glycemic response

5.3.1 *Starchy Flours*

Starchy flours are traditionally used in the preparation of different foods. They include flours from cereals and pseudocereals such as wheat, maize, rice, and amaranth; legumes such as beans, chickpeas, faba beans, peas, and lentils; tubers and roots such as potato, sweet potato, and taro; and unripe fruits such as banana, plantain, and mango. Native starchy flours such as wheat, maize, rice, potato, and sweet potato are traditionally used in bakery products, snacks, pasta, and cookies; they are mixed with other ingredients, such as eggs, butter, oil, sugar, and maltodextrins; and processed by baking, extrusion, frying, steaming, and boiling. They influence starch gelatinization and digestive enzyme accessibility to produce starch digestion. The use of unconventional starchy flour is an alternative to prepare gluten-free foods with low GI. Some of those unconventional starchy flours can present high dietary fiber content and a specific starch structure that reduces the starch digestion rate. For example, two types of unripe plantain flour were prepared, one from pulp and the other from pulp and peel (whole). It is well known that starch of raw unripe plantain is highly resistant to hydrolysis by digestive enzymes, but the cooking methods (boiling with peel and baking) can produce partial disorganization of starch that shows slow digestion and resistance to hydrolysis by digestive enzymes. Semolina was blended with unripe plantain flour at three different concentrations (15, 30, and 45% w/w), and the RS content in the cooked pasta increased from 2.84% at 15% of substitution to 12.42% at 45% of substitution, while the control pasta showed 1.1% of RS content [29]. The use of isolated banana starch as an ingredient in semolina pasta at different levels (5, 10, 15, and 20% w/w) presented increased RS content from 3.06% at 5% of substitution to 10.33% at 20% of substitution [30]. The use of isolated banana starch or banana flour influenced starch digestion in cooked pasta. The whole unripe plantain flour was blended with semolina at 15, 25, and 35% w/w. The digestible starch fractions were determined in cooked samples in a boiling water bath. Both RDS and SDS fractions showed a slight increase and a decrease, respectively, along with an increase in the substitution level and RS content from 0.8 (15% substitution) to 6.5% (35% substitution) [31].

Whole starchy flours are used to produce foods with increased dietary fiber content. The role of dietary fiber in human health is well known, and for the past decades, the development of foods containing dietary fiber has continued as an alternative to battle overweight and obesity as well as those diseases associated with them, such as diabetes, cardiovascular disease, and cancer [32]. Whole flour (wheat, maize, sorghum, and quinoa) is used to make bread, pasta, cookies, breakfast cereals, bars, and snacks commercially available. However, the global whole-grain intake is well below the recommended amounts [33]. It is important to know the major food sources of whole grains worldwide to design and implement strategies that increase the consumption of whole-grain foods [34].

Non-starch polysaccharides, components of dietary fiber, decrease the starch hydrolysis rate, contributing to satiety and regulating levels of blood sugar and cholesterol. Those effects are due to their ability to increase the viscosity of chyme and digesta, producing a physical barrier for digestive enzymes to hydrolyze the macromolecules present in the food [35]. It has been suggested that the presence of dietary fiber components (non-starch polysaccharides) influences starch digestion rate in maize starch with the addition of cellulose, where α-amylase is inhibited by the polysaccharide [13]. Future trends in research on the role of dietary fiber components in foods are to explore complex systems like whole foods and by-products with high dietary fiber content (fruit peels, grain husk), and evaluate both dietary fiber components (soluble and insoluble) [35].

5.3.2 High Amylose Starches

Amylose is an essential, linear component of the starch granule, and its molar mass and size are lower than those of amylopectin. During the thermal treatment before consumption of starchy foods with water, amylose lixiviates from the granule toward the continuous phase. Upon cooling, amylose chains are arranged in a network, which is more compact and rigid at longer times, in a process called retrogradation. The retrograde structure is not a substrate for digestive enzymes. Wild-type starch contains between 15% and 35% amylose. The increase in amylose content in the granule modifies polymorphism, birefringence, and size and shape. It has been reported that high-amylose maize starch (HAMS) shows elongated granules with two or more birefringent zones [36].

High-amylose starch and flour (Table 5.3) are proposed as a strategy to increase the indigestible starch content. High-amylose maize, barley, and potato are commercially available, and more recently, other food crops as wheat and rice, have

Table 5.3 Digestibility of starch-based ingredients with high-amylose content

Source	Starch digestibility comments	References
High-Amylose starch	The residues of the unfermented starch largely remained unchanged during the fermentation process. The wild-type starch with 37% amylose content showed a preferential degradation of the amylopectin fraction	[37]
High-Amylose flour	The slow digestion and resistant starch fraction were higher than in the wild-type flour (37% amylose)	[38]

been modified to increase the amylose content. Different strategies have been proposed to generate high-amylose starches by engineering the starch biosynthesis pathways: (1) suppressing starch synthases to decrease amylopectin content; (2) suppressing starch branching enzymes; and (3) suppressing debranching enzymes [39].

The mechanisms to increase RS in high-amylose starch are based on restricting granule swelling during heating (cooking). The restriction does not completely disorganize the starch components in the granule; intact granules remain (RS2), and they are not hydrolyzed by digestive enzymes. Additionally, during the heating, the high amylose content in the granule produces lixiviation of more amylose chains. During cooling, a network is produced, reducing accessibility of digestive enzymes, and the structure of the network is not recognized by the digestive enzymes (RS3). It is claimed that adding commercially available HAMS (Hi-Maize 260®) to different foods increases the indigestible carbohydrate content (dietary fiber), and at least 60% of the added ingredient is maintained as dietary fiber. Lipids and polyphenols in the food matrix can produce complexes with the linear amylose chains lixiviated during cooking (RS5). High-amylose wheat flour (soft) (HAWF) has been used to prepare bread [38], noodles [40], and tortillas [41] (Table 5.4). Bread baked with HAWF showed a lower loaf volume than that made with regular wheat flour, and its RS content was 11–12 times higher than that of regular wheat bread [38]. The noodles prepared with HAWF showed higher food integrity, producing a physical barrier against starch hydrolysis by digestive enzymes. The RS content was eightfold higher than in the noodles made using regular wheat flour [40]. Additionally, the use of different HAWF concentrations has been extended to tortillas. Some drawbacks of HAWF tortillas are the lower breaking force and extensibility due to the incomplete starch gelatinization, yet this reduced starch digestion while RS increased from 3% in the control (32% amylose) to 12% in HAWF tortilla with 71% amylose and 23% in HAWF tortillas with 84% amylose. The local density of starch microstructure formed during food processing affects the starch hydrolysis rate [43].

There is a growing interest in increasing the amylose content in different starchy food crops with additional strategies, as supplying soil with nutrients. The role of amylose in different food matrices has been evaluated, the major one being to reduce the starch hydrolysis rate. The use in other food matrices and thermal processes deserves more investigation.

Table 5.4 High-amylose starch and flour are used for the preparation of foodstuffs

Source	Product	Starch digestibility comments	References
High-amylose rice starch	Gluten-free fettuccine	This study produced Gluten-free fettuccine with a low glycemic index by replacing rice flour with a starch-lipid complex prepared by heat-inducing a high amylose rice starch to form a complex with stearic acid. The Gluten-free fettuccine prepared by replacing rice flour with starch-lipid complex at 20% had higher slow digestible starch (20.87%) and resistant starch (4.34%) content and lower eGI value (67.06) than Gluten-free fettuccine without starch-lipid complex addition (91.49)	[42]
High-amylose wheat flour	Noodle	The resistant starch in the high-amylose wheat flour was eight-fold higher than normal wheat flour (wild-type)	[40]
High-amylose wheat flour	Bread	The starch present in the bread prepared with high-amylose flour retained its crystalline structure compared with the wild-type flour bread. This molecular characteristic was related to the higher resistance to hydrolysis by digestive enzymes. High-amylose wheat bread showed 12-fold higher resistant starch than wild-type wheat bread	[38]
High-amylose wheat flour	Tortilla	The starch digestibility of tortillas elaborated with the high-amylose wheat flour was lower than with the wild-type. The resistant starch content increased (12 and 23%) when high-amylose wheat flour with 71% and 84% were used for tortilla preparation, respectively	[41]

Table 5.5 Modified starch and flour used in foodstuff elaboration

Source	Modification type	Product	Starch digestibility comments	References
Banana starch	Citric acid	Gluten-free cookies	The optimal condition to produce cookies with high resistant starch, low pGI and glycemic load (GL) were baking temperature of 158 °C and baking time of 20 min with predicted resistant starch value of 19.61 g/100 g of dry starch, pGI value of 56.98 and GL value 52.64	[44]
Banana starch	Hydroxypropylation	Pasta	The GI blended pasta changed significantly when compared with control and Commercial pasta. Control and commercial pastas showed comparatively higher GI value (87.9 and 91.5) than the pasta samples with added green banana flour and banana-modified starch (83.0–77.2)	[45]
Tapioca starch	Chemically modified tapioca starch	Bread	Chemically modified tapioca produced a bread formulation with a low predicted glycemic index (pGI < 55). Retrograded starch from tapioca allows the bread to reach a pGI value of 55, the upper value for classifying a food as low pGI. The retrograded starch from corn allows a decrease in the bread's glycemic index, but the product is still classified as 'high pGI' (>70)	[46]

5.3.3 *Modified Starch and Flour*

Modifying starchy flour and isolated starches has been a strategy to reduce the starch digestion rate and the glycemic response of foods. There is a recent interest in using eco-friendly methods for starch modification with reduced negative effects on the environment (Table 5.5).

Starch modification to decrease hydrolysis by digestive enzymes aims to change the arrangement of the double helices of amylopectin chains and, consequently, the crystallinity. The debranching of starch (amylopectin) with enzymes (isoamylase and pullulanase) to produce linear chains, followed by heating-cooling cycles, produces an arrangement of those linear chains in a retrograded structure that is resistant to hydrolysis by digestive enzymes. Debranching time can be determined to obtain a structure of linear chains to produce SDS and RS. It has been reported that high debranching enzyme concentration and shorter times (3–6 h) increase SDS content, and increased debranching time accelerates RS production [22]. The combination of enzymatic treatment (debranching) and physical methods (heating-cooling) modifies the starch structure and starch digestion rate. The chemical modification of starch has been decreased, involving the reagents that introduce functional groups in starch molecules (glucose monomers) [47]. Chemical modifications such as hydroxypropylation, cross-linking, esterification, acetylation, oxidation, and a combination of two or more of them have been successfully used to reduce starch digestion by increasing SDS and RS [48–51]. The chemical modifications of starch have some drawbacks, such as the generation of residues that pollute the environment, while the use of the reagents is regulated by the Federal Drug Administration (FDA) because changes in starch chains are restricted. Gluten-free breads show a rapid staling that increases the hardiness of the product and reduces its shelf-life. Commercial chemically modified starches have been proposed to reduce this problem. Di-starch phosphate, acetylated di-starch adipate, and pregelatinized acetylated di-starch phosphate have been evaluated with substitutions of 10% and 20% in breads. The study focused on evaluating the textural properties of the breads and the evolution of starch retrogradation through moisture loss and reassociation of amylopectin chains [52]. The chemical modification had a greater effect on the evaluated variables than extrusion did. However, the impact in the glycemic response of the substituted gluten-free breads has not been evaluated.

Enzymatic approaches to structuring starch are of interest to the food industry because they reduce undesirable byproducts and coproducts, and are not regulated by the FDA. These enzymatic methods are considered environmentally friendly to produce clean-label starches with functional and physicochemical attributes [53]. The combination of β-amylase and transglucosidase to hydrolyze starch increased the number of short chains and an increased α-1,6 linkages that produce a slow digestion starch rate [54]. The dual modification of unconventional starches (plantain and mango) using β-amylase and transglucosidase produced an increase in plantain SDS from 7.15% in β-amylase to 18.53% in dually modified starch. In contrast, an inverse pattern was observed in RS content (18.05 and 32.74%, respectively). Dually modified mango starch showed lower SDS content than its β-amylase counterpart, and the opposite was observed in RS content. It is possible to define the type of enzymatic treatment to modify the starch structure and produce SDS and RS [55].

The modification of starchy flour through proteins, lipids, non-starch polysaccharides, and bioactive compounds should be considered before selecting the method to modify starch structure and those macromolecules forming the food matrix that restrict hydrolysis by digestive enzymes. Starchy flour modification employs physical methods as hydrothermal treatments and extrusion. Those treatments alter the structure and, consequently, the functionality of the main components (starch, proteins, lipids, and non-starch polysaccharides). Extrusion has been used to obtain pregelatinized flours that can be used as a fat replacer and an ingredient in gluten-free foods. Amaranth flour was pregelatinized (49% gelatinization) and used in gluten-free and low-fat cakes. The control formulation contained 29 g of amaranth flour and 14 g of sunflower oil. The oil was replaced with 4.47, 9.33, and 14 g of pregelatinized amaranth flour, corresponding to 33.3, 66.6, and 100% fat replacement, respectively. The increase in pregelatinized amaranth flour did not affect the RS content but reduced SDS and increased RDS at the expense of fat reduction. In our opinion, more studies on pregelatinized starchy flours and starch digestion rate in diverse food matrices deserve future research.

The modification of wheat flour with superheated steam can increase the gluten force under soft conditions because an increase in the intensity of the treatment could produce the denaturation of the proteins [56, 57]. Changes in the functional properties of wheat flour and cake quality after treatment were associated with protein solubility, breaking of disulfide bonds, molecular weight, and secondary protein structure (α-helix, β-sheet, and β-coil). In a strict sense, starch digestion rate and glycemic response are associated with the restricted access of digestive enzymes to starch due to protein arrangement and perhaps non-starch polysaccharides in the food matrix.

5.3.4 Amylose–Lipids Complex

The presence of internal lipids in the starch showed a complex formation with linear chains, mainly amylose. The complex formation and two endothermic transitions, indicative of non-equilibrium melting, were evidenced by differential scanning calorimetry (DSC) using a heating rate of 10 °C/min. This was explained by partial melting, followed by recrystallization and final melting. The presence of an amylose–lipid complex was observed in rice starch. A phase transition around 105–110 °C in rice starch in excess water was attributed to the complex because it is stable to temperatures above starch gelatinization. This thermal characteristic can be important when the complex is produced and added to a food matrix that requires a thermal treatment before consumption. The starch–lipid complex can be created during the thermal processing of a food matrix when starch and lipids are present. The linear component of starch (amylose and long-side amylopectin chains) is prone to producing a complex with fatty acids in lipids. Two main structures of the starch–lipid complex are reported. The V_I is a single helical structure produced by a hydrophobic interaction where the hydrophobic cavity of the linear chain of starch hosts the 12–18 carbon chain fatty acid. The V_{II}-type complex is generated *via* hydrophobic interactions (e.g., π interactions), which can be produced by increasing the molecular weight of the fatty acid or the presence of unsaturated fatty acids [9]. The advances in the structural studies of the starch–lipid complex led to the proposal of an update of the RS5 based on self-assembled starch V-type complexes [9]. The V-type complex was associated with its X-ray diffraction pattern and named arbitrarily as V-type to differentiate it from A-, B-, and C- type X-ray diffraction patterns [58].

The nutritional importance of the starch–lipid complex is that it is undigestible by digestive enzymes and is classified as RS5 [59]. This concept raised interest in the production of the complex as an ingredient in foods to increase the undigestible carbohydrate content, reducing glycemic response, and maintaining gut microflora due to fermentation in the colon. It is well known that the effect of RS consumption has been proposed for the intervention of insulin resistance and metabolic syndrome; this proposal was based on the reduction of postprandial plasma-glucose and insulin response [60, 61]. The starch–lipid complex with a crystalline arrangement presents a high melting temperature because it can be used in food matrices produced by thermal processing. The chain length and unsaturation of the fatty acids are important in the complex formation with varying degrees of order; monounsaturated fatty acids, as oleic acid, form a more orderly lamellar structure than polyunsaturated fatty acids (linoleic acid) [62]. The starch–lipid complex is produced during the thermal treatment of starchy foods with lipids. There are studies in model systems using steam jet cooking [63, 64] and heat-moisture treatment (HMT) with low moisture content and high temperature [65]. The starch–lipid complex was formed with oleic and linoleic acids, and the complex was different depending on the fatty acid. Linoleic acid produced a V_I-type structure, while oleic acid created a type II. The type depends on the melting temperature of the complex crystallinity; for instance, type I has a melting temperature below 100 °C, and type II, above 100 °C. This pattern is associated with the partially ordered structure with no distinct crystalline regions (Type I), and distinct crystalline and amorphous regions (Type II) [66]. The increase in fatty acid concentration decreased RDS, but when the same fatty acid concentration was used to prepare the complex, oleic acid showed a higher decrease in RDS than linoleic acid did. The RS content presented an inverse pattern: It increased compared with native rice starch and HMT rice starch. However, RS with the same fatty acid concentration was higher with oleic acid than with linoleic acid [65]. This issue was not clarified in the study and deserves more investigation. The production of the starch-lipid complex using this green method (HMT) is an alternative to produce ingredients with reduced glycemic response in the design of functional starchy food matrices.

The starch–lipid complex can be produced with high yield when employing high-amylose starch. It is well known that linear chains are prone to producing a complex with fatty acids. The starch–lipid complex was produced with HAMS and palmitic acid (PA) (saturated with 16 carbons). The complex produced with HAMS and PA produced 39.2% of RS, which was higher than in the control sample (without PA). Starch debranching with isoamylase to increase linear chain content increased the RS content (46.7%) due to starch retrogradation. The complex produced with the debranched starch showed the highest RS content (52.7%) due to starch retrogradation and complex formation [67].

5.3.5 *Amylose–Bioactive Compounds Complex*

Bioactive compounds are usually found in vegetable sources, such as fruits, roots, leaves, flowers, and peels, among others. They include phenolic acids, polyphenols, flavonoids, and anthocyanidins. Their role in human health includes modulation of oxidative and inflammatory stress, stimulation of the immune system, alteration of macronutrient digestion, and a prebiotic-like effect on the gut microbiota. In addition, evidence shows that the consumption of foods rich in polyphenols can reduce the risk of age-related chronic diseases [68]; the presence of polyphenols during processing of a food matrix may alter stability, bioaccessibility, bioavailability, and fermentation in the colon [69]. Diverse studies have been published on the interactions of polyphenols and starch; the linear glucan chains can join to the polyphenols with the production of V_I-type, V_{IIa}-type, and V_{IIb}-type complexes, generated depending on the polyphenol type (structure) and the development of a food matrix [70–72]. The complexes are resistant to hydrolysis by the digestive enzymes. If they are released from the complex, polyphenols can inhibit amylases and decrease the starch hydrolysis rate in the small intestine. The complex can continue the journey toward the colon, where it is a substrate for the microbiota [73]. One strategy is the encapsulation of bioactive compounds using different wall materials, as maltodextrins and cyclodextrins, by spray-drying and freeze-drying [74]. Freeze-dried samples show a higher encapsulation efficiency of green tea polyphenols than spray-drying. The encapsulation with maltodextrins retains a higher antioxidant capacity [75].

The production of complex polyphenols and starch by non-covalent V-complexes is of particular interest due to the polyphenols being encapsulated for controlled delivery. These complexes can be produced and used as ingredients to provide nutraceutical characteristics to the food matrix. Polyphenols are hosted in the hydrophobic helical cavity of linear glucan chains and may act as carriers for bioactive compounds, which could be released upon hydrolysis of the glucan chain [76]. The production of complexes involves the formation of the hydrophobic cavity in linear glucan chains, mainly high-amylose starch, and the next step is the addition of bioactive compounds. The effect of chain lengths of alkyl gallates on complex formation using amylose or high-amylose starch has been evaluated. Gallic acid and alkyl gallates with short chains did not form inclusion complexes with both linear glucan chain sources; gallates with longer alkyl chains were able to form more stable inclusion complexes. The gallates with long alkyl chains showed the lowest RDS content and highest SDS and RS content [77]. Different bioactive compounds have been complexed with linear starch chains, such as proanthocyanidins [78], isoflavones [79], and curcumin [80]. Complex formations between starch and bioactive compounds are an alternative to produce a nutraceutical ingredient with low glycemic response since starch is partially or not completely hydrolyzed in the small intestine, and the released host compound can inhibit digestive enzymes and increase non-digestible starch. The antioxidant properties of bioactive compounds can produce health benefits widely reported. The bioavailability of bioactive compounds and the bioactive compounds released after fermentation by the microbiota must be further investigated.

5.4 Food Applications

The starch–lipid complex with the highest RS content was used as an ingredient in bread preparation. The bread showed an RS content of 34.4% because the consumption of 100 g of this bread meets the requirements of the dietary fiber recommendation. The glucose in the plasma of nonsmoking men was substantially reduced compared to when subjects consumed white bread [67]. The addition of phenolic compounds to cooked porridges (120 °C for 30 min) prepared with starch and the blend of starch/protein was analyzed under different environmental conditions. These model systems provide information on the role of the different macromolecules included in a food matrix. The processing of the porridge produced an interaction between starch and polyphenols to make the complex. The study analyzed the recovery of phenolic compounds after in vitro digestion. Phenolic compounds are unstable through digestion in the intestinal phase due to the elevated pH (6.5–7.0). It was found that phenolic compounds are less degraded when starch is only present in the porridge than in the presence of both macromolecules (starch and protein) [81]. The effect of berry polyphenols on starch digestion rate was evaluated in bread fortified with berry extract and the co-digestion of bread and berry extract. Fortified bread showed a higher extent of starch hydrolysis than the co-digestion; this pattern was because in the food matrix of the fortifying, the bioaccessibility of polyphenols was reduced, reducing the amount of polyphenols available to inhibition for α-amylase [82]. The interactions between starch and polyphenols during the co-digestion of starchy food and polyphenol extract is a promising strategy to reduce GI.

5.5 Conclusions

Using starchy food ingredients to reduce the glycemic index has been related to a decrease in the starch digestion rate. Strategies for reduction include the use of high-amylose starch and interactions (complex formation) between macromolecules such as starch–lipid, starch–proteins, and starch–bioactive compounds. Those interactions are produced during the processing of the food matrix and the co-digestion of the starchy food matrix and bioactive compounds extracts. The interactions between starch and lipids have been proposed to produce a complex and used as an ingredient in a food matrix. However, the production of this complex to expand its use deserves further investigation. The complex between starch and bioactive compounds has been studied to understand the mechanisms involved in its production. The basic studies of that complex include starch hydrolysis rate and the release of bioactive compounds. However, its use as a functional ingredient in the starchy food matrix is yet to be reported.

Acknowledgments The authors gratefully acknowledge the Instituto Politécnico Nacional, CEPROBI for the support.

Conflicts of Interest The authors declare no conflict of interest.

References

1. Kesre C, Masatcioglu MT. Physical characteristics of corn extrudates supplemented with red lentil bran. LWT. 2022;153(112530):112530. https://doi.org/10.1016/j.lwt.2021.112530
2. Miao M, Hamaker BR. Food matrix effects for modulating starch bioavailability. Annu Rev Food Sci Technol. 2021;12(1):169–91. https://doi.org/10.1146/annurev-food-070620-013937
3. Grand View Research, Inc. Healthy snacks market size, share & trends analysis report, 2023 – 2030. San Francisco: Grand View Research, Inc.; 2021.
4. Grand View Research, Inc. North America healthy snack chips market size, share & trends analysis report by product (tortilla chips, pop chips, popcorn chips, puff chips, potato chips, tapioca chips, and others), by country, and segment forecasts, 2024 – 2030. 2025. Grandviewresearch.com. https://www.grandviewresearch.com/industry-analysis/north-america-healthy-snack-chips-market
5. Grand View Research, Inc. Global bakery & confectionery industrial starch market statistics, 2018–2030 (USD Million). 2025. Grandviewresearch.com. https://www.grandviewresearch.com/horizon/statistics/bakery-confectionery-global-industrial-starch-market-906690
6. Grand View Research, Inc. U.S. biscuits market size, share & trends analysis report by type, by product (cookies, filled/coated biscuits, crackers, cream wafers), by source, by distribution channel, by region, and segment forecasts, 2025 – 2030. 2025. Grandviewresearch.com. https://www.grandviewresearch.com/industry-analysis/us-biscuits-market.
7. Grand View Research, Inc. Pulse ingredients market size, share & trends analysis report by source (peas, chickpea, lentils), by type (pulse flour, pulse starch), by application (food & beverage, feed), by region, and segment forecasts, 2024 – 2030. 2025. Grandviewresearch.com. https://www.grandviewresearch.com/industry-analysis/pulse-ingredients-market-report
8. Grand View Research, Inc. Resistant starch market size, share & trends analysis report by source (vegetables, grains, cereal food), by product (RS1, RS2), by application (beverages, confectionery, dairy products), by region, and segment forecasts, 2023 – 2030. 2025. Grandviewresearch.com. https://www.grandviewresearch.com/industry-analysis/resistant-starch-market
9. Gutiérrez TJ, Tovar J. Update of the concept of type 5 resistant starch (RS5): self-assembled starch V-type complexes. Trends Food Sci Technol. 2021;109:711–24. https://doi.org/10.1016/j.tifs.2021.01.078.
10. Xu H, Zhang G. Slow digestion property of microencapsulated normal corn starch. J Cereal Sci. 2014;60(1):99–104. https://doi.org/10.1016/j.jcs.2014.01.021.
11. Yang W, Hattori M, Kawaguchi T, Takahashi K. Properties of starches conjugated with lysine and poly(lysine) by the Maillard reaction. J Agric Food Chem. 1998;46(2):442–5. https://doi.org/10.1021/jf970515i.
12. Granfeldt Y, Björck I. Glycemic response to starch in pasta: a study of mechanisms of limited enzyme availability. J Cereal Sci. 1991;14(1):47–61. https://doi.org/10.1016/s0733-5210(09)80017-9
13. Dhital S, Gidley MJ, Warren FJ. Inhibition of α-amylase activity by cellulose: kinetic analysis and nutritional implications. Carbohydr Polym. 2015;123:305–12. https://doi.org/10.1016/j.carbpol.2015.01.039
14. He T, Zhao L, Wang L, Liu L, Liu X, Dhital S, Hu Z, Wang K. Gallic acid forms V-amylose complex structure with starch through hydrophobic interaction. Int J Biol Macromol. 2024;269:129408
15. Bello-Perez LA, Flores-Silva PC, Agama-Acevedo E, Tovar J. Starch digestibility: past, present, and future: starch digestibility: past, present, and future. J Sci Food Agric. 2020;100(14):5009–16. https://doi.org/10.1002/jsfa.8955
16. Jenkins DJ, Wolever TM, Taylor RH, Barker H, Fielden H, Baldwin JM, et al. Glycemic index of foods: a physiological basis for carbohydrate exchange. Am J Clin Nutr. 1981;34(3):362–6. https://doi.org/10.1093/ajcn/34.3.362

17. Miao M, Jiang B, Cui SW, Zhang T, Jin Z. Slowly digestible starch-a review. Crit Rev Food Sci Nutr. 2015;55:1642–57.
18. Goñi I, Garcia-Alonso A, Saura-Calixto F. A starch hydrolysis procedure to estimate glycemic index. Nutr Res. 1997;17(3):427–37. https://doi.org/10.1016/s0271-5317(97)00010-9.
19. Englyst HN, Kingman SM, Cummings JH. Classification and measurement of nutritionally important starch fractions. Eur J Clin Nutr. 1992;46 Suppl 2:S33–50.
20. Cantu-Jungles TM, Hamaker BR. New view on dietary fiber selection for predictable shifts in gut microbiota. MBio. 2020;11(1):e02179–19. https://doi.org/10.1128/mBio.02179-19.
21. Brouns F, Kettlitz B, Arrigoni E. Resistant starch and "the butyrate revolution". Trends Food Sci Technol. 2002;13(8):251–61. https://doi.org/10.1016/s0924-2244(02)00131-0.
22. Miao M, Jiang B, Zhang T. Effect of pullulanase debranching and recrystallization on structure and digestibility of waxy maize starch. Carbohydr Polym. 2009;76(2):214–21. https://doi.org/10.1016/j.carbpol.2008.10.007.
23. Zhang G, Hasek LY, Lee B-H, Hamaker BR. Gut feedback mechanisms and food intake: a physiological approach to slow carbohydrate bioavailability. Food Funct. 2015;6(4):1072–89. https://doi.org/10.1039/c4fo00803k.
24. Magallanes-Cruz PA, Flores-Silva PC, Bello-Perez LA. Starch structure influences its digestibility: a review: structure-digestibility relationship. J Food Sci [Internet]. 2017;82(9):2016–23. Disponible en: 10.1111/1750-3841.13809
25. Zhang G, Ao Z, Hamaker BR. Slow digestion property of native cereal starches. Biomacromolecules. 2006;7(11):3252–8. https://doi.org/10.1021/bm060342i.
26. Zhang G, Venkatachalam M, Hamaker BR. Structural basis for the slow digestion property of native cereal starches. Biomacromolecules. 2006;7(11):3259–66. https://doi.org/10.1021/bm060343a0.
27. Butterworth PJ, Warren FJ, Grassby T, Patel H, Ellis PR. Analysis of starch amylolysis using plots for first-order kinetics. Carbohydr Polym. 2012;87(3):2189–97. https://doi.org/10.1016/j.carbpol.2011.10.048.
28. Bello-Perez LA, Agama-Acevedo E, Garcia-Valle DE, Alvarez-Ramirez J. A multiscale kinetics model for the analysis of starch amylolysis. Int J Biol Macromol. 2019;122:405–9. https://doi.org/10.1016/j.ijbiomac.2018.10.161.
29. Ovando-Martinez M, Sáyago-Ayerdi S, Agama-Acevedo E, Goñi I, Bello-Pérez LA. Unripe banana flour as an ingredient to increase the undigestible carbohydrates of pasta. Food Chem. 2009;113(1):121–6. https://doi.org/10.1016/j.foodchem.2008.07.035.
30. Hernández-Nava RG, Berrios JDJ, Pan J, Osorio-Díaz P, Bello-Perez LA. Development and characterization of spaghetti with high resistant starch content supplemented with banana starch. Food Sci Technol Int. 2009;15(1):73–8. https://doi.org/10.1177/1082013208102379.
31. Garcia-Valle DE, Agama-Acevedo E, Alvarez-Ramirez J, Bello-Perez LA. Semolina pasta replaced with whole unripe plantain flour: chemical, cooking quality, texture, and starch digestibility. Star. 2020;72(9–10):1900097. https://doi.org/10.1002/star.201900097.
32. Carlson JL, Erickson JM, Lloyd BB, Slavin JL. Health effects and sources of prebiotic dietary fiber. Curr Dev Nutr. 2018;2(3):nzy005. https://doi.org/10.1093/cdn/nzy005.
33. Odonovan CB, Devlin NF, Buffini M, Walton J, Flynn A, Gibney MJ, et al. Whole grain intakes in Irish adults: findings from the National Adults Nutrition Survey (NANS). Eur J Nutr. 2019;58:541–50.
34. Price EJ, Barrett EM, Batterham MJ, Beck EJ. Exploring the reporting, intake, and recommendations of primary food sources of whole grains globally: a scoping review. Br J Nutr. 2024;132(10):1365–76.
35. Karim A, Raji Z, Habibi Y, Khalloufi S. A review on the hydration properties of dietary fibers derived from food waste and their interactions with other ingredients: opportunities and challenges for their application in the food industry. Crit Rev Food Sci Nutr. 2024;64(32):11722–56. https://doi.org/10.1080/10408398.2023.2243510.

36. Cai J, Man J, Huang J, Liu Q, Wei W, Wei C. Relationship between structure and functional properties of normal rice starches with different amylose contents. Carbohydr Polym. 2015;125:35–44. https://doi.org/10.1016/j.carbpol.2015.02.067.
37. Li H, Gilbert RG, Gidley MJ. Molecular-structure evolution during in vitro fermentation of granular high-amylose wheat starch is different to in vitro digestion. Food Chem. 2021;362(130188):130188. https://doi.org/10.1016/j.foodchem.2021.130188.
38. Li C, Dhital S, Gidley MJ. High-amylose wheat bread with reduced in vitro digestion rate and enhanced resistant starch content. Food Hydrocoll. 2022;123(107181):107181. https://doi.org/10.1016/j.foodhyd.2021.107181.
39. Zhong Y, Qu JZ, Liu X, Ding L, Liu Y, Bertoft E, et al. Different genetic strategies to generate high amylose starch mutants by engineering the starch biosynthetic pathways. Carbohydr Polym. 2022;287(119327):119327. https://doi.org/10.1016/j.carbpol.2022.119327.
40. Li H-T, Li Z, Fox GP, Gidley MJ, Dhital S. Protein-starch matrix plays a key role in enzymic digestion of high-amylose wheat noodle. Food Chem. 2021;336(127719):127719. https://doi.org/10.1016/j.foodchem.2020.127719.
41. Li C, Dhital S, Gidley MJ. High-amylose wheat tortillas with reduced in vitro digestion and enhanced resistant starch content. Food Hydrocoll. 2023;137(108321):108321. https://doi.org/10.1016/j.foodhyd.2022.108321.
42. Fitriati A, Detchewa P, Moongngarm A. Replacing rice flour with starch-lipid complex reduced estimated glycemic index and enhanced slow digestible starch and resistant starch content in gluten-free fettuccine. J Cereal Sci. 2025;121(104080):104080. https://doi.org/10.1016/j.jcs.2024.104080.
43. Li C, Dhital S, Gidley MJ. High amylose wheat foods: a new opportunity to improve human health. Trends Food Sci Technol. 2023;135:93–101. https://doi.org/10.1016/j.tifs.2023.03.017.
44. Olawoye B, Gbadamosi SO, Otemuyiwa IO, Akanbi CT. Gluten-free cookies with low glycemic index and glycemic load: optimization of the process variables via response surface methodology and artificial neural network. Heliyon. 2020;6(10):e05117. https://doi.org/10.1016/j.heliyon.2020.e05117.
45. Keran DA, Kumar PS, Pushpavalli S, Uma S. In-vitro digestibility, textural and quality characteristics of ditalini pasta fortified with green banana flour and its type-IV modified starch. Int J Food Sci Technol. 2022;57(7):4656–72.
46. Pasqualoni I, Tolve R, Simonato B, Bianchi F. The impact of selected ingredients on the predicted glycemic index and technological properties of bread. Food Secur. 2024;13(16):2488. https://doi.org/10.3390/foods13162488.
47. Sandhu K, Singh N. Some properties of corn starches II: physicochemical, gelatinization, retrogradation, pasting and gel textural properties. Food Chem. 2007;101(4):1499–507. https://doi.org/10.1016/j.foodchem.2006.01.060.
48. Han J-A, BeMiller JN. Preparation and physical characteristics of slowly digesting modified food starches. Carbohydr Polym. 2007;67(3):366–74. https://doi.org/10.1016/j.carbpol.2006.06.011.
49. Carlos-Amaya F, Osorio-Diaz P, Agama-Acevedo E, Yee-Madeira H, Bello-Pérez LA. Physicochemical and digestibility properties of double-modified banana (Musa paradisiaca L.) starches. J Agric Food Chem. 2011;59(4):1376–82.
50. Simsek S, Ovando-Martínez M, Whitney K, Bello-Pérez LA. Effect of acetylation, oxidation and annealing on physicochemical properties of bean starch. Food Chem. 2012;134(4):1796803. https://doi.org/10.1016/j.foodchem.2012.03.078.
51. Simsek S, Ovando-Martinez M, Marefati A, Sjöö M, Rayner M. Chemical composition, digestibility and emulsification properties of octenyl succinic esters of various starches. Food Res Int. 2015;75:41–9. https://doi.org/10.1016/j.foodres.2015.05.034.
52. Roman L, Yee J, Hayes AMR, Hamaker BR, Bertoft E, Martinez MM. On the role of the internal chain length distribution of amylopectins during retrogradation: double helix lateral aggregation and slow digestibility. Carbohydr Polym. 2020;246(116633):116633. https://doi.org/10.1016/j.carbpol.2020.116633.

53. Miao M, BeMiller JN. Enzymatic approaches for structuring starch to improve functionality. Annu Rev Food Sci Technol. 2023;14(1):271–95. https://doi.org/10.1146/annurev-food-072122-023510
54. Miao M, Li R, Huang C, Jiang B, Zhang T. Impact of β-amylase degradation on properties of sugary maize soluble starch particles. Food Chem. 2015;177:1–7. https://doi.org/10.1016/j.foodchem.2014.12.101
55. Casarrubias-Castillo MG, Hamaker BR, Rodriguez-Ambriz SL, Bello-Pérez LA. Physicochemical, structural, and digestibility properties of enzymatic modified plantain and mango starches. Star. 2012;64(4):304–12. https://doi.org/10.1002/star.201100142
56. Hu Y, Wang L, Zhu H, Li Z. Modification of physicochemical properties and in vitro digestibility of wheat flour through superheated steam processing. J Cereal Sci. 2017;74:231–7. https://doi.org/10.1016/j.jcs.2017.02.012
57. Ma Y, Sang S, Xu D, Jin Y, Chen Y, Xu X. The contribution of superheated steam treatment of wheat flour to the cake quality. LWT. 2021;141:110958
58. Zobel HF. Molecules to granules: a comprehensive starch review. Star. 1988;40(2):44–50. https://doi.org/10.1002/star.19880400203
59. Jane J-L, Robyt JF. Structure studies of amylose-V complexes and retro-graded amylose by action of alpha amylases, and a new method for preparing amylodextrins. Carbohydr Res. 1984;132(1):105–18. https://doi.org/10.1016/0008-6215(84)85068-5
60. Jenkins DJ, Wolever TM, Jenkins AL. Starchy foods and glycemic index. Diabetes Care. 1988;11(2):149–59. https://doi.org/10.2337/diacare.11.2.149.
61. Englyst KN, Englyst HN, Hudson GJ, Cole TJ, Cummings JH. Rapidly available glucose in foods: an in vitro measurement that reflects the glycemic response. Am J Clin Nutr. 1999;69(3):448–54. https://doi.org/10.1093/ajcn/69.3.448
62. Ai Y, Hasjim J, Jane J-L. Effects of lipids on enzymatic hydrolysis and physical properties of starch. Carbohydr Polym. 2013;92(1):120–7. https://doi.org/10.1016/j.carbpol.2012.08.092
63. Byars JA, Fanta GF, Felker FC. Rheological properties of dispersions of spherulites from jet-cooked high-amylose corn starch and fatty acids. Cereal Chem. 2009;86(1):76–81. https://doi.org/10.1094/cchem-86-1-0076
64. Fanta GF, Kenar JA, Byars JA, Felker FC, Shogren RL. Properties of aqueous dispersions of amylose–sodium palmitate complexes prepared by steam jet cooking. Carbohydr Polym. 2010;81(3):645–51. https://doi.org/10.1016/j.carbpol.2010.03.026
65. He H, Zheng B, Wang H, Li X, Chen L. Insights into the multi-scale structure and in vitro digestibility changes of rice starcholeic acid/linoleic acid complex induced by heat-moisture treatment. Food Res Int. 2020;137(109612):109612. https://doi.org/10.1016/j.foodres.2020.109612
66. Obiro WC, Sinha Ray S, Emmambux MN. V-amylose structural characteristics, methods of preparation, significance, and potential applications. Food Rev Int. 2012;28(4):412–38. https://doi.org/10.1080/87559129.2012.6607187
67. Hasjim J, Lee S-O, Hendrich S, Setiawan S, Ai Y, Jane J-L. Characterization of a novel resistant-starch and its effects on postprandial plasma-glucose and insulin responses. Cereal Chem. 2010;87(4):257–62. https://doi.org/10.1094/cchem-87-4-0257
68. Shahidi F, Yeo J. Bioactivities of phenolics by focusing on suppression of chronic diseases: a review. Int J Mol Sci. 2018;19(6):1573. https://doi.org/10.3390/ijms19061573
69. Debelo H, Li M, Ferruzzi MG. Processing influences on food polyphenol profiles and biological activity. Curr Opin Food Sci. 2020;32:90–102. https://doi.org/10.1016/j.cofs.2020.03.00
70. Romero-Hernández HA, Gutiérrez TJ, Bello-Pérez LA. Can starch-polyphenol V-type complexes be considered as resistant starch? Food Hydrocoll. 2022;124:107226
71. Gutiérrez TJ, Bello-Pérez LA. Self-assembled and assembled starch V-type complexes for the development of functional foodstuffs: a review. Food Hydrocoll. 2022;125(107453):107453. https://doi.org/10.1016/j.foodhyd.2021.107453
72. Hernández-Bautista M, Heredia-Olea E, Agama-Acevedo E, Bello-Pérez LA. Extrusion of decoction residues of Roselle (Hibiscus sabdariffa L.) with different starchy wall

materials retaining the polyphenol content after in vitro digestion. Food Chem. 2024;467(142322):142322. https://doi.org/10.1016/j.foodchem.2024.142322
73. Marín L, Miguélez EM, Villar CJ, Lombó F. Bioavailability of dietary polyphenols and gut microbiota metabolism: antimicrobial properties. Biomed Res Int. 2015;2015:905215. https://doi.org/10.1155/2015/905215
74. Papillo VA, Locatelli M, Travaglia F, Bordiga M, Garino C, Coïsson JD, et al. Cocoa hulls polyphenols stabilized by microencapsulation as functional ingredient for bakery applications. Food Res Int. 2019;115:511–8. https://doi.org/10.1016/j.foodres.2018.10.004
75. Pasrija D, Ezhilarasi PN, Indrani D, Anandharamakrishnan C. Microencapsulation of green tea polyphenols and its effect on incorporated bread quality. Lebenson Wiss Technol. 2015;64(1):28996. https://doi.org/10.1016/j.lwt.2015.05.054
76. Zhu Y-F, Chen J-J, Ji X-M, Hu X, Ling T-J, Zhang Z-Z, et al. Changes of major tea polyphenols and production of four new B-ring fission metabolites of catechins from post-fermented Jing-Wei Fu brick tea. Food Chem. 2015;170:110–7. https://doi.org/10.1016/j.foodchem.2014.08.075
77. Zhang Y, Gladden I, Guo J, Tan L, Kong L. Enzymatic digestion of amylose and high amylose maize starch inclusion complexes with alkyl gallates. Food Hydrocoll. 2020;108:106009. https://doi.org/10.1016/j.foodhyd.2020.106009
78. Amoako DB, Awika JM. Resistant starch formation through intrahelical V-complexes between polymeric proanthocyanidins and amylose. Food Chem. 2019;285:326–33. https://doi.org/10.1016/j.foodchem.2019.01.173
79. Wang S, Wu T, Cui W, Liu M, Wu Y, Zhao C. Structure band in vitro digestibility on complex of corn starch with soy isoflavone. Food Sci Nutr. 2020;8:6061–8
80. Sun L, Warren FJ, Gidley MJ. Soluble polysaccharides reduce binding and inhibitory activity of tea polyphenols against porcine pancreatic α-amylase. Food Hydrocoll. 2018;79:63–70. https://doi.org/10.1016/j.foodhyd.2017.12.011
81. Ferruzzi MG, Hamaker BR, Bordenave N. Phenolic compounds are less degraded in presence of starch than in presence of proteins through processing in model porridges. Food Chem. 2020;309(125769):125769. https://doi.org/10.1016/j.foodchem.2019.125769
82. Kan L, Oliviero T, Verkerk R, Fogliano V, Capuano E. Interaction of bread and berry polyphenols affects starch digestibility and polyphenols bio-accessibility. J Funct Foods. 2020;68(103924):103924. https://doi.org/10.1016/j.jff.2020.103924

Chapter 6
Starch Structuration and Digestibility

J. Alvarez-Ramirez and E. J. Vernon-Carter

6.1 Introduction

Proteins are essential Starch is the only polysaccharide of vegetal origin that can be digested by humans. The incorporation of glucose in human metabolism is achieved via enzymatic hydrolysis reactions to generate glucose, which can be easily absorbed in the digestive tract. In this way, starch is the most accessible source of energy for human activity. Structurally, starch is composed of two polysaccharides with different chemical structures. Amylose is a linear polymer of glucose units linked by α-(1–4) bonds, in which some α-(1–6) bonds may be present. On the other hand, amylopectin contains branches that give it a branched molecular shape, where the branches are linked to the central trunk by α-D-(1, 6) bonds, located every 25–30 linear glucose units [1].

Starch as a source of energy has accompanied the development of civilization since ancient times. Cereals have been by far the main botanical sources of starch in human history [2]. Ancient cultures flourished thanks to their skills in domesticating, producing, storing, and distributing cereals. Wheat is one of the oldest and most cultivated cereals, originating in the Near East. The domestication of wheat occurred around 10,000 BC, when humans became sedentary and focused on developing agriculture [3]. Wheat spread throughout Asia and Europe, driving the development of ancient civilizations, including Mesopotamia, Egypt, Greece, and Rome. Wheat has been consumed mainly in the form of bread and, to a lesser extent, in beverages, desserts, and sweets. Rice is another highly consumed cereal, and its origin is in Southeast Asia. Rice is believed to have been domesticated from the wild species *Oryza rufipogon*, which grows in the swampy and lacustrine areas of the region [4].

J. Alvarez-Ramirez (✉) · E. J. Vernon-Carter
Departamento de Ingeniería de Procesos e Hidráulica, Universidad Autónoma Metropolitana-Iztapalapa, Iztapalapa, México
e-mail: jjar@xanum.uam.mx; jvc@xanum.uam.mx

G. Ayala Valencia (ed.), *Starch: Progress in Food Applications*,
https://doi.org/10.1007/978-3-031-98732-8_6

The domestication of rice is estimated to have occurred around 8000 years ago, during the Neolithic period. Rice played an important role in the development of many cultures in Asia, including China and India, and was introduced to America by the Spanish and Portuguese in the sixteenth century. Maize played a primordial role in the evolution of the peoples of America, including the Incas, Aztecs, and Mayans. Maize is believed to have been domesticated from the wild species Teosinte (*Zea mays* subsp. parvifolia), which grows in the mountainous and valley areas of Mesoamerica [5]. The domestication of maize is estimated to have occurred around 9000 years ago, during the Neolithic period. Maize consumption has spread globally, along with wheat and rice, as the most widely extended crops. Rye, oats, barley, and sorghum are other cereals whose consumption has been growing in recent centuries. Potato (*Solanum tuberosum*) is the fourth most important food crop in the world in terms of production volume. It was first domesticated in the region of modern-day southern Peru and extreme northwestern Bolivia about 7000 to 10,000 years ago. Potato became the basic crop of Europe in the eighteenth century. Potato played an important role in the industrialization of Europe, contributing to the large-scale industrialization of foods [6].

Historically, starch intake has been carried out mainly through its content in cereals, but also in tubers, legumes, and fruits. Wheat starch consumption has been predominantly through the preparation of baked goods. Wheat is consumed mainly through the preparation of dishes, although also through bread. Maize offers great diversity for its incorporation into traditional foods, tortillas, and snacks. Potatoes are widely consumed as a side dish in a variety of dishes in gastronomy and fast food. Oats have also contributed to starch intake through the preparation of cookies, snacks, and directly in breakfasts. In summary, the consumption of foods high in starch remains the most important, and often affordable, way to incorporate energy sources for human activity.

6.2 Historical Antecedents

6.2.1 Starch Consumption in Preindustrial Societies

Agriculture and livestock were the main sources of income and employment in pre-industrial societies, prior to the Industrial Revolution [7]. The production of food and other basic products such as wool, silk, and leather was essential for survival. With the family and society being the basic units of social organization, they were organized to cultivate the land, care for livestock, and perform other essential tasks. In this way, pre-industrial societies were organized around physical labor in a way that reflected their dependence on agriculture and livestock. Social organization, division of labor, and the use of simple tools and technologies were fundamental to the survival and success of these societies. The stability of society strongly depended on access to cereals, which provided the energy supplies necessary to maintain various productive, political, and social activities. Basically, most of the population in pre-industrial societies lived in conditions of energy balance, where food consumption provided sufficient energy, albeit in times of deficit, to carry out productive and social activities. In this scenario, cereals and their main component, starch, were the main agents of human nutrition.

6.2.2 *Starch Consumption in Industrialized Societies*

The growth of the population in Great Britain during the eighteenth century created an increase in the demand for goods and services, as well as the development of new transportation, supply- chains and good manufacturing processes [8]. The prevailing political stability, combined with growing external trade, allowed for the creation of a favorable environment for investment and economic growth. Agriculture experienced significant improvements, increasing food productivity and releasing labor for other activities. The combination of a series of almost fortuitous factors resulted in the development of new productive techniques based on innovative tools and machinery, such as the steam engine, mechanical spinners, and railways. The once physically demanding work was transformed into more operative work, where machinery partially or totally replaced human physical effort. The nineteenth century witnessed the dissemination of this new form of production, which was consolidated in the twentieth century with major technological developments in mechanization, automation, and digitization. While in pre-industrial societies, most of the population was rural, the industrial boom generated a high demand for organized labor around industrial centers. By the late twentieth century, the world population had become predominantly suburban and urban, leaving a reduced fraction in the rural environment.

The industrialization of social life led to people having more economic resources and leisure time. These factors resulted in a modification of lifestyle, where food was generally more abundant and diverse. Even in the pre-automation and digitization stage, production processes still required a relatively high expenditure of human energy. Assembly line machinery, urban construction, and administrative work depended heavily on human physical activity. In line with this, food production entered an industrialized phase where the goal was to have quick access to highly digestible foods. The invention of instant breakfast cereals in the early twentieth century marked the emergence of an industrial sector that rapidly changed the eating habits of Western societies. Instant cereals offered rapidly digestible starches, which should be associated with a healthy and active lifestyle.

6.2.3 Starch Consumption in Automatized/Digitalized Societies

With the advent of digitalization and automation starting in the 1980s, productive processes increasingly demanded less human physical intervention. The mechanization of agriculture had a significant impact on the displacement of rural populations to industrialized regions. Manufacturing industry workers were displaced by highly efficient robotic systems, while construction industry employees were partially replaced by semi-automatic machinery. Although the process of accelerated industrialization and automation developed in Western economies, it quickly spread to almost all regions of the world starting in the 1990s. While primary productive processes are based on a relatively small fraction of the population, a new employment sector emerged related to administration, financial and communication services, culture, tourism, and leisure. This growing sector involves most of the population and is characterized by low demand for human physical labor. Under increasingly strict regulations, the idea is that workers have relatively moderate working hours and controlled office or home working conditions, especially since the Covid-19 era. As a consequence, people spend most of their working hours at rest in front of digitalized equipment. The emergence of social networks and virtual entertainment environments increased sedentary behavior among most of the population, surprisingly including those living in relatively small towns.

The changes in production methods at the end of the twentieth century and the beginning of the twenty-first century, which led to more leisure time and increases in life expectancy, as well as advances in medicine and pharmacy, should lead to healthy individuals. The increased production of food, mainly cereals, due to the automation of crops and the improvement of seeds, generated a strong supply of relatively cheap processed foods. The sedentary lifestyle, combined with the consumption of foods rich in starches and lipids, triggered unprecedented health problems. While in the pre-industrial era and to a lesser extent in the industrial era, the energy consumed in food was balanced with the metabolic energy expenditure in physical work, the current pattern is characterized by a positive balance in energy intake. The "Dietary Guidelines for Americans" developed by the USDA in 1980 recommended a low-fat diet rich in carbohydrates. However, based on limited evidence, it led to a significant increase in the consumption of refined flour and added sugars, contributing to the rise in obesity and overweight. In 1995, the World Health Organization (WHO) recognized the impending problem and published a report defining obesity as a chronic and multifactorial disease, with adverse effects on cardiovascular health. In 2000, the "Healthy People 2010" report from the US Department of Health and Human Services set specific goals to reduce the prevalence of obesity and overweight in the US population.

Although the structuring of starch for reduced digestibility is a central research topic, the development of starchy foods for specialized population sectors is also of interest. Populations suffering from food deficiencies in various regions of the world require highly digestible starch. The performance of endurance athletes depends on the rapid availability of dietary energy, for which carbohydrates must be structured to dose glucose absorption [9]. The case of elderly people is also relevant, as digestive issues are commonly affected by gradual reduction in digestive enzyme production [10].

6.3 Metabolic Syndrome

The prevalence of overweight and obesity was accompanied by significant health problems, including cardiac dysfunction, hypertension, and diabetes. These health derivatives were grouped into a single framework called metabolic syndrome [11, 12]. Recent studies have shown a population prevalence of 20% of metabolic syndrome [13, 14]. Such levels of prevalence lead to recognizing metabolic syndrome as a public health problem in many regions of the world [15]. Although sedentary lifestyle is an important factor in the incidence of metabolic syndrome, dietary patterns play a central role in the incidence of the problem.

The amount and type of fats consumed in food were initially explored as the central cause of obesity and overweight [16, 17]. Soon, it was recognized that the problem was multifactorial in terms of human diet [18]. Various reports showed that carbohydrates made a significant contribution to the development of metabolic syndrome [19, 20]. Excess glucose in the energy balance can result in health risks associated with metabolic syndrome. In this way, excessive sugar consumption was an obvious factor in plasma glucose levels [21]. Since starch is a polysaccharide, whose enzymatic hydrolysis generates glucose in the digestive tract and given the high starch content in a diet based on processed foods, it was recognized as an important contributor to the growth of problems associated with metabolic syndrome [22]. The change in productive systems from demanding human physical force to one where force is exercised by automatic systems resulted in sedentary lifestyles, without adapting dietary patterns to a more energetically balanced one.

6.4 Starch Digestion

There is a consensus that starch plays a central role in the incidence of health problems related to metabolic syndrome [23]. Starch chains are basically glucose, which, once released by enzymatic reactions, is available for absorption in the digestive tract. Therefore, understanding starch digestion and glucose metabolism is crucial for designing strategies to reduce the negative impact of starch intake. Roughly, starch digestion begins in the mouth with chewing and the release of amylolytic enzymes to produce maltose [24, 25]. Starch hydrolysis at this stage is relatively low. The chyme's stay in the stomach hardly involves enzymatic hydrolysis since α-amylase is rapidly irreversibly inhibited at pH's of around 3. Digestive juices (hydrochloric acid, potassium chloride, and sodium chloride) and stomach muscle movements reduce the particle size of the chyme. By effect of gastric emptying, the chyme is transported to the small intestine, where the pancreas releases amylolytic enzymes that convert starch into maltotriose, maltose, limit dextrins, and some glucose. Enzymes located on the luminal surfaces of enterocytes lining the microvilli of the small intestine convert limit dextrins and disaccharides into glucose, galactose, and fructose, which are absorbed through the epithelial cells of the small intestine via the sodium-potassium pump and driven by two separate transport proteins (SGLT1 and GLUT2) [26]. The starch hydrolysis process in the small intestine is accompanied by the transport, via peristalsis, of the chyme along approximately 4–5 meters [27]. The process ends when the chyme is discharged into the large intestine, where the microbiota takes advantage of the residues from enzymatic digestion to produce a series of metabolites. Once in the blood plasma, glucose can be transported to the mitochondria to proceed with the energy cycle, incorporated into the glycogen cycle in the liver, or converted into lipids through de novo lipogenesis in fat stored in adipocytes [28].

6.5 Starch Digestibility

The availability of glucose for absorption in the small intestine greatly depends on the kinetics of enzymatic hydrolysis of starch. The pioneering work of Jenkins et al. [29] highlighted that hydrolysis can be carried out through mechanisms acting at different scales, which can be exploited to regulate glucose generation in the small intestine. The packing of starch chains into granules is an obstacle to the access of amylolytic enzymes to the binding sites. In this way, the long-range structure of the starch granule could be manipulated to modulate glucose availability in the small intestine. On the other hand, the fine lamellar and double-helix structure becomes important when the integrity of the starch granule is affected. The fine arrangement of the starch chains could be related to slowing the rate of hydrolysis. However, measurements of in vivo enzymatic hydrolysis of starch are costly and difficult to perform, even currently. Englyst et al. [30] proposed an in vitro test to evaluate the kinetics of hydrolysis and, in this way, estimate the impact of starch content on the glycemic index of a food. These authors classified starch into three fractions depending on the rate of enzymatic hydrolysis. Rapidly digestible starch (RDS) is that which is hydrolyzed in the first 20 minutes, while slowly digestible starch (SDS) corresponds to the fraction that is hydrolyzed between 20 and 120 min. The starch remaining after 120 min is considered resistant starch (RS). The maximum limit of 120 min was assumed to be the average time that starch remains in the small intestinal tract. Goñi et al. [31] further contributed to the topic by proposing a simple model to estimate the rate of starch hydrolysis.

The recognition that starch can contribute to problems associated with metabolic syndrome and the work of Englyst et al. [30] opened a new avenue in food research. From the 1990s onwards, the paradigm has been to establish the mechanisms of starch digestion and how these mechanisms can be manipulated, via structural modification, to modulate the rate of enzymatic hydrolysis. For this purpose, in vitro analysis of hydrolysis kinetics has become the most widely used method due to its easy implementation and low costs. The recent three decades have seen the publication of a plethora of results on starch digestibility. The effect of granule morphology, relative contents of amylose and amylopectin, molecular structure, and degree of amylopectin branching are factors that have been widely explored [32]. The role of starch resisting enzymatic hydrolysis has been increasingly understood in terms of its effects on glucose production in the small intestine and its impact on the gut microbiota [33]. The importance of slowly digestible starch in the regulation of glucose availability in the small intestine, and therefore in blood plasma, has also been recognized. The benefits of slowly digestible starch are related to adequate diabetes management, satiety, and mental performance [34, 35].

On the other hand, resistant starch acts as dietary fiber, exerting several benefits on the functioning of the digestive system, including reduction of fat accumulation, reduction of insulin levels, and modulation of blood pressure. Additionally, the multiscale structure of resistant starch is exploited by gut microbiota to generate functional compounds (e.g., vitamins) that, when absorbed, have a positive impact on human health [36]. Recent studies have shown evidence that consuming resistant starch aids in weight loss by reconfiguring the population composition of the gut microbiota [37].

6.6 Methods of Starch Modification

6.6.1 Physical Methods

Commonly, native starches present a high degree of enzymatic hydrolysis, with values as high as 60–70% in granules and 90–100% in gelatinized dispersions [38]. In this way, many research efforts have focused on exploring methods to modify the structure of starches to reduce the rate and maximum degree of enzymatic hydrolysis. One objective has been the development of starches with a high content of slowly digestible and resistant starches.

Physical methods aim to modify the packing of starch chains without affecting the D-glucopyranosyl units, which impacts the properties of starch, including pasting and gel properties, and even digestibility [39]. The attractiveness of physical methods lies in the fact that their application does not require the intervention of chemical agents, making such modifications generally safe for human consumption. Additionally, physical methods are considered environmentally friendly alternatives due to the low or null generation of toxic waste [40].

Heat-moisture treatment is an approach where starch is subjected to heating at temperatures above the glass transition temperature. The implementation of the treatment is relatively simple, involving only the adjustment of the moisture content. Several studies reviewed by da Rosa Zavareze and Días [41] reported that hydrothermal treatment impacts the amorphous regions of starch granules and generates depolymerization, resulting in increased susceptibility to amylolytic reactions. In this way, modifications based on hydrothermal treatment can be considered suitable for obtaining starch with a high degree of digestibility.

Microwave heating (300–300,000 MHz) induced by electromagnetic fields is another treatment that has been explored. The moisture content and treatment duration are two of the main variables adjusted to induce uniform heating. Regarding the effects of the treatment on digestibility, some results point towards increased susceptibility [42], while others indicate a notable increase in the content of slowly digestible and resistant starch.

Ultrasonic treatments (16–18 kHz) have found acceptance for modifying the structure of starch granules. Ultrasonication produces cavitation in the liquid medium, which, upon collapse, produces macrozones of high shear stress accompanied by intense local heating and rapid cooling [43]. It has been found that ultrasonic modification produces starch with a high degree of digestibility. The effect was attributed to the formation of short-chained amylose molecules by effects of ultrasonic cavitation [44].

High hydrostatic pressure (> 400 MPa) improves thermal stability, reduces the gelatinization temperature, and alters the crystallinity of starch granules. Some studies found that this type of treatment can reduce the content of rapidly digestible starch [45, 46]. A disadvantage of the treatment is that it requires highly specialized equipment, which can lead to a notable increase in production costs.

Pulsed electric field treatment has been explored as an alternative to modify the structure of starch granules with minimal denaturation effects. The impact of the treatment is not well characterized, with results pointing towards both positive and negative effects depending on the botanical source [47]. However, the results have shown that the strategy based on electrical pulses has great application potential to modify the fine structure of starch granules. On the other hand, the method does not require the intermediation of a chemical agent, which makes it an environmentally friendly procedure.

6.6.2 Chemical Methods

Physical methods do not alter the chemical structure (D-glucopyranosyl units) of starch chains. This can be considered a limitation to the scope of modifications and their effects on starch functional properties. Additionally, physical methods generally lead to increased digestibility, which is not a desirable effect for managing health issues related to, e.g., metabolic syndrome. Chemically-based methods, on the other hand, directly impact on the chemical structure of biopolymers and commonly aim to reduce the rate of enzymatic hydrolysis [48].

Oxidation methods affect the three hydroxyl groups, which are substituted through serial reactions. Starch oxidation can be carried out using sodium hypochlorite, hydrogen peroxide, ozone, and sodium periodate. Oxidation modifies the functional properties of starch, leading to a reduction in gel viscosity, changes in crystalline structure, and thermal stability [49]. Oxidative processes decrease digestibility, with an increase in slow-digesting and resistant starch contents [50]. Dry heating at temperatures of 140–160 °C induces oxidation of starch chains by the action of air oxygen, resulting in a decrease in relative crystallinity, an increase in low-range order (double-helix structures) of the chains. of starch and a significant increase in the content of resistant starch [51]. In contrast to oxidation methods that use a chemical oxidizing agent, dry heating can be considered an environmentally friendly method without traces of toxicity that could be harmful to human health.

Starch esterification involves the conversion of hydroxyl groups into alkyl or aryl derivatives. Acetylation is the most commonly used approach to induce starch esterification. Native starch is activated with alkaline, resulting in the formation of an alkoxide, which reacts with a hydroxide group to form an acetylated derivative [52]. Octenyl succinic anhydride (OSA) is an esterification agent whose use for starch modification has grown in recent decades [53]. Gels of esterified starch exhibit low viscosity and crystallinity, making these derivatives suitable for use in the food industry as thickeners and stabilizers. However, esterification can negatively affect starch digestibility, leading to the formation of resistant starch [54, 55].

Etherification involves the substitution of hydroxyl groups with carboxymethyl, hydroxypropyl, or hydroxyethyl groups through the formation of an ether link. In general, the etherification reaction is catalyzed with sodium hydroxide [56]. Gels of etherified starch exhibit improved thermal stability and flowability. Cross-linking and acid hydrolysis are other methods that have been explored for modifying starch functionality [57]. The impact on digestibility is varied, with some studies reporting reduced susceptibility to enzymatic hydrolysis [58].

Overall, chemically based methods result in profound modifications to the chemical structure of starch chains. Commonly, such modifications are associated with reduced digestibility, reflected by a decrease in rapidly digestible starch. However, there are concerns among consumers about potential health risks that may arise from the use of chemical modifying agents [59].

Chemical methods have been combined with physical methods to increase the efficiency of the action of the chemical agent on the structure of starch chains. The underlying idea is to generate a synergistic effect where the modification of starch is carried out by a chemical agent, while a physical action (e.g., ultrasound, microwave) acts as a promoter of the transport of reactants and heat generator on a local scale. Esterification catalyzed with Lipozyme 435 lipase can be improved by ultrasound intervention, resulting in a restructuring of the starch chains requiring a reduced amount of catalyst [60]. In a similar vein, the use of low-intensity ultrasound (up to 20 W/cm^2) promoted esterification with octenyl succinic anhydride (OSA) of annatto seed starch [61]. The combination of ultrasound and ozone induced the modification of the morphology, mechanical, thermal and barrier properties of corn starch [62]. The combination of microwaves with other methods has been used to obtain a synergistic effect in the modification of the starch structure. The results are promising since the combination of methods confers better functionality and commonly reduced digestibility than that with simple treatments [63]. Additionally, the combination of chemical and physical methods can reduce the requirement for chemical reagents to achieve the desired modification (e.g., degree of substitution).

6.6.3 Starch-Lipid-Protein Complexes

The interaction between lipids and starch results in the formation of complexes with various applications in the food industry. Their importance lies in their ability to influence the texture and nutritional quality of food products. Starch-lipid complexes can be used as fat replacers, stabilizers, and anti-staling agents. An interesting property is that the starch involved in these complexes resists hydrolysis by amylolytic enzymes (RS Type 5) [64]. However, the degree of resistance depends on the botanical source and molecular structure of the starch [65]. Starch and proteins can interact to form complexes, which are commonly used as emulsifying agents and texture stabilizers in food [66]. The formation of complexes with proteins increases the content of slowly digestible and resistant starch [67]. Triple interactions between starch, lipids, and proteins have been studied in recent years, with results showing that starch digestibility is markedly reduced [68]. The study of starch-based complexes is important because, in addition to being formulated synthetically, they are an intrinsic part of many food products. Therefore, characterizing starch-based complexes is crucial for understanding and characterizing the nutritional and functional properties of a food product [69].

6.6.4 Genetic Modifications

Genetic manipulation of seeds is a common practice in recent decades. The underlying idea is to obtain crops that offer higher productivity, pest resistance, and expression of desirable traits (e.g., higher protein content). The genetic modification method is controversial and subject to discussions to establish strict regulations for the consumption of these products by humans [70]. Modern technologies such as CRISPR have been explored to produce rice with high amylose content and resistant starch [71]. More general results have pointed out in the direction of producing cereals with low digestibility through genetic manipulation of seeds [72]. Given the increase in the production and consumption of cereals from genetically modified seeds, and the growing concern on the part of consumers, detailed studies on the impact on the digestibility and safety of flour and starch are required.

6.6.5 *Starch-Polyphenols Complexes*

Starch chains can interact with polyphenols to form V-type complexes [73]. Various studies have found that polyphenols can reduce starch digestibility by inhibiting the action of amylolytic enzymes [74]. Other studies have found that the presence of tannins and proanthocyanidins increases the content of slowly digestible starch and, in some cases, resistant starch [75, 76]. It has been found that the formation of complexes and the decrease in starch digestibility are correlated with the number of hydroxyl groups of polyphenols [77]. Additionally, it has been found that the effect of polyphenols on starch structure is multiscale, impacting both crystallinity and the ordering of short-range structures (e.g., double helices) [78]. Most studies have focused on the synthesis of phenol-mediated complexes. In many cases, polyphenols are an inherent component of foods, such as whole grains, pigmented maize, fruits, legumes, and vegetables. Apparently, these food sources already contain polyphenol-starch complexes, which modulate the extent of digestibility. The ultra-processing of foods can affect the stability of such complexes, resulting in an increase in starch digestibility. Thus, research efforts should be devoted to characterizing the impact of food processing methods on the stability and digestibility of starch bound to antioxidant compounds.

6.7 Mathematical Models of Starch Digestion

Mathematical modeling of the kinetics of enzymatic hydrolysis of starch is important for quantifying starch fractions (e.g., RDS and SDS) and for helping to elucidate the factors that affect digestibility. Goñi et al. [31] proposed an empirical model given by a first-order exponential to describe digestograms. Fitting experimental data provides an estimate of the rate (i.e., hydrolysis rate constant) of enzymatic hydrolysis. Although the model is simple, it is by far the most widely used for modeling digestograms. Modifications to the first-order model have been proposed to obtain a better fit of experimental data. Second-order exponentials to account for multiscale structures [79], extensions of the Michaelis-Menten model [80], and Weibull functions [81] are models that have shown good data fitting capabilities. Models based on log-of-slope (LOS) analysis have been established as a useful tool for detecting transitions in the kinetics of enzymatic starch digestion [82]. The LOS method is simple and the information obtained from the analysis has been useful for undercovering the impact of starch structure and modification on starch digestion mechanisms.

Monte Carlo simulations have been explored to gain insights into the transport mechanisms and interactions between amylolytic enzymes and starch in granules and gelatinized dispersions [83]. The results suggested that enzyme transport within the starch granule structure can lead to multiscale hydrolysis, which is reflected in the presence of slowly and rapidly digestible starches. On the other hand, the resistant starch commonly reported in most studies is not truly resistant, but rather results from the inability to complete hydrolysis within the average residence time (about 120 min) in the large intestine. Wen et al. [36] reported that RS with different fine structures and properties is consumed to varying degrees by gut microbiota, producing metabolites in the small intestine.

Some efforts have been dedicated to developing mathematical models of starch digestion in the digestive tract. Experimentally, in vitro models based on biological membranes have been a substantial advance compared to simple tests [84]. Semipermeable membranes are often used, which act as a selective barrier to keep the enzymes separated from their digestion products, avoiding the inhibition of enzymatic hydrolysis. The information derived from these tests is useful for the development of mathematical models that take into account the physicochemical and transport phenomena in the small intestine. In such a framework, models of enzymatic hydrolysis kinetics are used as constitutive equations to evaluate the impact of starch structure on glucose availability. Meraz et al. [27] developed a model given by a partial differential equation with a moving boundary to describe the transport of chyme and the generation of glucose by enzymatic hydrolysis. Numerical analysis of the model showed that hydrolysis kinetics has a central impact on blood glucose homeostasis. Qin et al. [85] used a transport model to evaluate the effect of mass transfer on glucose absorption in the small intestine. The development of glucose generation models in the small intestine highlights the importance of considering not only hydrolysis kinetics but also the mechanisms (e.g., mass transfer, intestinal motility) involved in glucose absorption.

The mathematical models commented above have focused on the hydrolysis and availability of starch in the small intestine. An important aspect is the metabolism of residual starch that was not hydrolyzed and is transferred to the large intestine. Depending on the structure and physicochemical conditions in the small intestine, the values of undigested starch can vary in the range 5–40%. Undigested starch, even those with high resistance, is fermented by the microbiota to produce a series of metabolites, such as vitamins and short-chain fatty acids [86]. Some efforts have been devoted to developing mathematical models to describe carbohydrate fermentation by human colonic microbiota [87]. It has been pointed out that mathematical models in this direction could be of great value in the rational design of microbiota-tailoring diets and therapies. In particular, the analysis of the impact of starch structure on gut microbiota dynamics could be analyzed [88].

6.8 Conclusions

Research on starch digestibility and its relationship to human nutrition and health has experienced strong growth in recent decades. The results reported in the specialized literature have uncovered mechanisms and factors that importantly determine digestibility patterns. However, the analysis and digestibility of starch in food science and technology are far from being resolved. The recognition of health problems associated with metabolic syndrome, the emergence of increasingly critical consumers, and the diversification of nutritional needs open up a broad avenue for research on starch digestibility. The majority of studies are limited to reporting the most basic aspects of digestibility, including hydrolysis rate constants and the distribution of digestible fractions. Although this information may be relevant, it is insufficient to obtain a realistic evaluation of how the structure and nature of native and modified starches impact glucose availability in the digestive tract [89]. Improved in vivo, in vitro, and in silico models of digestibility must be developed and established as standard methodologies in evaluating digestibility. In this way, the food science and technology researcher and developer must be not only a researcher but also a perceptive critic of the technological developments available on the market. The future research on starches in the food industry and human nutrition is broad, promising innovative results and new multidisciplinary perspectives [90]. The modification of the starch structure, as well as the interaction with other food components (e.g., lipids, proteins, and polyphenols) are topics that should be at the center of the analysis.

Acknowledgements The authors acknowledge the support given by the Division de Ciencias Básicas e Ingeniería of Universidad Autónoma Metropolitana Campus Iztapalapa for the development of this study.

Conflicts of Interest The authors declare no conflict of interest.

References

1. Whistler RL, BeMiller JN, Paschall EF, editors. Starch: chemistry and technology. 2nd ed. Academic Press; 2012.
2. Hoegemeyer T, Francis C, Baenziger S. Our daily bread: a history of cereals. Crop Sci. 2014;54(1):453.
3. Goncharov NP, Golovnina KA, Kilian B, Glushkov S, Blinov A, Shumny VK. Evolutionary history of wheats—the main cereal of mankind. In: Biosphere origin and evolution. Springer; 2008. p. 407–19.
4. Sweeney M, McCouch S. The complex history of the domestication of rice. Ann Bot. 2007;100(5):951–7.
5. Goodman MM, Galinat WC. The history and evolution of maize. Crit Rev Plant Sci. 1988;7(3):197–220.
6. Salaman RN, Burton WG. The history and social influence of the potato. Cambridge University Press; 1985.
7. Sellen DW, Smay DB. Relationship between subsistence and age at weaning in "preindustrial" societies. Hum Nat. 2001;12:47–87.
8. De Vries J. The industrial revolution and the industrious revolution. J Econ Hist. 1994;54(2):249–70.
9. Ormsbee MJ, Bach CW, Baur DA. Pre-exercise nutrition: the role of macronutrients, modified starches and supplements on metabolism and endurance performance. Nutrients. 2014;6(5):1782–808.
10. Pérez Sira EE. Foods for the elderly. In: Foods for special dietary regimens. Bentham Books; 2021. p. 58–78.

11. Eckel RH, Grundy SM, Zimmet PZ. The metabolic syndrome. Lancet. 2005;365(9468): 1415–28.
12. Alberti KGM, Zimmet P, Shaw J. The metabolic syndrome—a new worldwide definition. Lancet. 2005;366(9491):1059–62.
13. Manaf MRA, Nawi AM, Tauhid NM, Othman H, Rahman MRA, Yusoff HM, Safian N, Ng PY, Manaf ZA, Kadir NBA, Yanasegaran K, Basir SMA, Ramakrishnappa S, Ganasegeran K. Prevalence of metabolic syndrome and its associated risk factors among staffs in a Malaysian Public University. Sci Rep. 2021;11(1):8132.
14. Ali N, Samadder M, Shourove JH, Taher A, Islam F. Prevalence and factors associated with metabolic syndrome in university students and academic staff in Bangladesh. Sci Rep. 2023;13(1):19912.
15. Nilsson PM, Tuomilehto J, Rydén L. The metabolic syndrome–what is it and how should it be managed? Eur J Prevent Cardiol. 2019;26(2_suppl):33–46.
16. Rothwell NJ, Stock MJ. The development of obesity in animals: the role of dietary factors. Clin Endocrinol Metab. 1984;13(3):437–49.
17. Applegate EA, Upton DE, Stern JS. Exercise and detraining: effect on food intake, adiposity and lipogenesis in osborne-mendel rats made obese by a high fat diet. J Nutr. 1984;114:447–59.
18. Valavanis IK, Mougiakakou SG, Grimald KA, Nikita KS. A multifactorial analysis of obesity as CVD risk factor: use of neural network based methods in a nutrigenetics context. BMC Bioinform. 2010;11:1–10.
19. Wylie-Rosett J, Segal-Isaacson CJ, Segal-Isaacson A. Carbohydrates and increases in obesity: does the type of carbohydrate make a difference? Obes Res. 2004;12(S11):124S–9S.
20. Van Dam RM, Seidell JC. Carbohydrate intake and obesity. Eur J Clin Nutr. 2007;61(1):S75–99.
21. Sundborn G, Thornley S, Merriman TR, Lang B, King C, Lanaspa MA, Johnson RJ. Are liquid sugars different from solid sugar in their ability to cause metabolic syndrome? Obesity. 2019;27(6):879–87.
22. Tapsell LC. Diet and metabolic syndrome: where does resistant starch fit in? J AOAC Int. 2004;87(3):756–60.
23. Svihus B, Hervik AK. Digestion and metabolic fates of starch, and its relation to major nutrition-related health problems: a review. Starch-Stärke. 2016;68(3–4):302–13.
24. McQuilken SA. Digestion and absorption. Anaesth Intensive Care Med. 2021;22(5):336–8.
25. Brownlee IA, Gill S, Wilcox MD, Pearson JP, Chater PI. Starch digestion in the upper gastrointestinal tract of humans. Starch-Stärke. 2018;70(9–10):1700111.
26. Yoshikawa T, Inoue R, Matsumoto M, Yajima T, Ushida K, Iwanaga T. Comparative expression of hexose transporters (SGLT1, GLUT1, GLUT2 and GLUT5) throughout the mouse gastrointestinal tract. Histochem Cell Biol. 2011;135:183–94.
27. Meraz M, Vernon-Carter EJ, Bello-Perez LA, Alvarez-Ramirez J. Mathematical modeling of gastrointestinal starch digestion-blood glucose-insulin interactions. Biomed Signal Process Control. 2022;77:103812.
28. Zierler K. Whole body glucose metabolism. Am J Physiol-Endocrinol Metab. 1999;276(3):E409–26.
29. Jenkins DJ, Wolever TM, Kalmusky J, Giudici S, Giordano C, Wong GS, Bird JN, Patten R, Hall M, Buckley G, Little JA. Low glycemic index carbohydrate foods in the management of hyperlipidemia. Am J Clin Nutr. 1985;42(4):604–17.
30. Englyst HN, Veenstra J, Hudson GJ. Measurement of rapidly available glucose (RAG) in plant foods: a potential in vitro predictor of the glycaemic response. Br J Nutr. 1996;75(3):327–37.
31. Goñi I, Garcia-Alonso A, Saura-Calixto F. A starch hydrolysis procedure to estimate glycemic index. Nutr Res. 1997;17(3):427–37.
32. Singh J, Dartois A, Kaur L. Starch digestibility in food matrix: a review. Trends Food Sci Technol. 2010;21(4):168–80.
33. Bello-Perez LA, Flores-Silva PC, Agama-Acevedo E, Tovar J. Starch digestibility: past, present, and future. J Sci Food Agric. 2020;100(14):5009–16.
34. Lehmann U, Robin F. Slowly digestible starch–its structure and health implications: a review. Trends Food Sci Technol. 2007;18(7):346–55.

35. Miao M, Jiang B, Cui SW, Zhang T, Jin Z. Slowly digestible starch—A review. Crit Rev Food Sci Nutr. 2015;55(12):1642–57.
36. Wen JJ, Li MZ, Hu JL, Tan HZ, Nie SP. Resistant starches and gut microbiota. Food Chem. 2022;387:132895.
37. Li H, Zhang L, Li J, Wu Q, Qian L, He J, Ni Y, Kovatcheva-Datchary P, Yuan R, Liu S, Shen L, Zhang M, Sheng B, Li P, Kang K, Wu L, Fang O, Long X, Wang X, Li Y, Ye Y, Ye J, Bao Y, Zhao Y, Xu G, Liu X, Panagiotou G, Xu A, Jia W. Resistant starch intake facilitates weight loss in humans by reshaping the gut microbiota. Nat Metab. 2024;6(3):578–97.
38. Corgneau M, Gaiani C, Petit J, Nikolova Y, Banon S, Ritié-Pertusa L, et al. Digestibility of common native starches with reference to starch granule size, shape and surface features towards guidelines for starch-containing food products. Int J Food Sci Technol. 2019;54(6):2132–40.
39. BeMiller JN, Huber KC. Physical modification of food starch functionalities. Annu Rev Food Sci Technol. 2015;6(1):19–69.
40. Maniglia BC, Castanha N, Le-Bail P, Le-Bail A, Augusto PE. Starch modification through environmentally friendly alternatives: a review. Crit Rev Food Sci Nutr. 2021;61(15):2482–505.
41. da Rosa ZE, Dias ARG. Impact of heat-moisture treatment and annealing in starches: a review. Carbohydr Polym. 2011;83(2):317–28.
42. Emami S, Perera A, Meda V, Tyler RT. Effect of microwave treatment on starch digestibility and physico-chemical properties of three barley types. Food Bioprocess Technol. 2012;5:2266–74.
43. Dutta D, Sit N. Ultrasound treatment. In: Starch. CRC Press; 2025. p. 132–64.
44. Flores-Silva PC, Roldan-Cruz CA, Chavez-Esquivel G, Vernon-Carter EJ, Bello-Perez LA, Alvarez-Ramirez J. In vitro digestibility of ultrasound-treated corn starch. Starch-Stärke. 2017;69(9–10):1700040.
45. Liu H, Fan H, Cao R, Blanchard C, Wang M. Physicochemical properties and in vitro digestibility of sorghum starch altered by high hydrostatic pressure. Int J Biol Macromol. 2016;92:753–60.
46. Zhang J, Zhang M, Bai X, Zhang Y, Wang C. The impact of high hydrostatic pressure treatment time on the structure, gelatinization and thermal properties and in vitro digestibility of oat starch. Grain Oil Sci Technol. 2022;5(1):1–12.
47. Li Q, Wu QY, Jiang W, Qian JY, Zhang L, Wu M, Rao S-Q, Wu CS. Effect of pulsed electric field on structural properties and digestibility of starches with different crystalline type in solid state. Carbohydr Polym. 2019;207:362–70.
48. Masina N, Choonara YE, Kumar P, du Toit LC, Govender M, Indermun S, Pillay V. A review of the chemical modification techniques of starch. Carbohydr Polym. 2017;157:1226–36.
49. Vanier NL, El Halal SLM, Dias ARG, da Rosa ZE. Molecular structure, functionality and applications of oxidized starches: a review. Food Chem. 2017;221:1546–59.
50. Zhou F, Liu Q, Zhang H, Chen Q, Kong B. Potato starch oxidation induced by sodium hypochlorite and its effect on functional properties and digestibility. Int J Biol Macromol. 2016;84:410–7.
51. Reyes I, Hernandez-Jaimes C, Vernon-Carter EJ, Bello-Perez LA, Alvarez-Ramirez J. Air oxidation of corn starch: effect of heating temperature on physicochemical properties and in vitro digestibility. Starch-Stärke. 2021;73(3–4):2000237.
52. Otache MA, Duru RU, Achugasim O, Abayeh OJ. Advances in the modification of starch via esterification for enhanced properties. J Polym Environ. 2021;29(5):1365–79.
53. Lopez-Silva M, Agama-Acevedo E, Bello-Perez LA, Alvarez-Ramirez J. Influence of gelatinization degree and octenyl succinic anhydride esterification on the water sorption characteristics of corn starch. Carbohydr Polym. 2021;270:118378.
54. Mei JQ, Zhou DN, Jin ZY, Xu XM, Chen HQ. Effects of citric acid esterification on digestibility, structural and physicochemical properties of cassava starch. Food Chem. 2015;187:378–84.
55. Xu T, Zhong Y, Chen Q, Wu L, Ji S, Yang B, Zhang Y, Shen J, Lu B. Modulating the digestibility of cassava starch by esterification with phenolic acids. Food Hydrocoll. 2022;127:107432.
56. Heinze T, Liebert T, Heinze U, Schwikal K. Starch derivatives of high degree of functionalization 9: carboxymethyl starches. Cellulose. 2004;11:239–45.

57. Ali TM, Haider S, Shaikh M, Butt NA, Zehra N. Chemical crosslinking, acid hydrolysis, oxidation, esterification, and etherification of starch. In: Advanced research in starch. Springer Nature Singapore; 2024. p. 47–94.
58. Ding L, Huang Q, Xiang W, Fu X, Zhang B, Wu JY. Chemical cross-linking reduces in vitro starch digestibility of cooked potato parenchyma cells. Food Hydrocoll. 2022;124:107297.
59. Gałkowska D, Kapuśniak K, Juszczak L. Chemically modified starches as food additives. Molecules. 2023;28(22):7543.
60. Vartolomei A, Calinescu I, Vinatoru M, Gavrila AI. A parameter study of ultrasound assisted enzymatic esterification. Sci Rep. 2022;12(1):1421.
61. Silva EK, Anthero AGDS, Emerick LB, Zabot GL, Hubinger MD, Meireles MAA. Low-frequency ultrasound-assisted esterification of Bixa orellana L. Seed Starch with Octenyl Succinic Anhydride. Int J Biol Macromol. 2022;207:1–8.
62. Huang X, Chen L, Liu Y. Effects of ultrasonic and ozone modification on the morphology, mechanical, thermal and barrier properties of corn starch films. Food Hydrocoll. 2024;147:109376.
63. Oyeyinka SA, Akintayo OA, Adebo OA, Kayitesi E, Njobeh PB. A review on the physicochemical properties of starches modified by microwave alone and in combination with other methods. Int J Biol Macromol. 2021;176:87–95.
64. Sengupta A, Chakraborty I, Mazumder N. An insight into the physicochemical characterisation of starch-lipid complex and its importance in food industry. Food Rev Intl. 2023;39(7):4198–212.
65. Sun S, Hong Y, Gu Z, Cheng L, Ban X, Li Z, Li C. Different starch varieties influence the complexing state and digestibility of the resulting starch-lipid complexes. Food Hydrocoll. 2023;141:108679.
66. Dong D, Cui B. Fabrication, characterization and emulsifying properties of potato starch/soy protein complexes in acidic conditions. Food Hydrocoll. 2021;115:106600.
67. Zhang Z, Zhang M, Zhao W. Effect of starch-protein interaction on regulating the digestibility of waxy rice starch under radio frequency treatment with added CaCl2. Int J Biol Macromol. 2023;232:123236.
68. Zhen Y, Wang K, Wang J, Qiao D, Zhao S, Lin Q, Zhang B. Increasing the pH value during thermal processing suppresses the starch digestion of the resulting starch-protein-lipid complexes. Carbohydr Polym. 2022;278:118931.
69. Yang Z, Zhang Y, Wu Y, Ouyang J. Factors influencing the starch digestibility of starchy foods: a review. Food Chem. 2023;406:135009.
70. Turnbull C, Lillemo M, Hvoslef-Eide TA. Global regulation of genetically modified crops amid the gene edited crop boom– a review. Front Plant Sci. 2021;12:630396.
71. Kumar A, Sahoo U, Lal MK, Tiwari RK, Lenka SK, Singh NR, Gupta OP, Sah RP, Sharma S. Biochemical markers for low glycemic index and approaches to alter starch digestibility in rice. J Cereal Sci. 2022;106:103501.
72. Yi X, Li C. Main controllers for improving the resistant starch content in cooked white rice. Food Hydrocoll. 2022;122:107083.
73. Hernández HAR, Gutiérrez TJ, Bello-Pérez LA. Can starch-polyphenol V-type complexes be considered as resistant starch? Food Hydrocoll. 2022;124:107226.
74. Miao M, Jiang H, Jiang B, Li Y, Cui SW, Zhang T. Structure elucidation of catechins for modulation of starch digestion. LWT-Food Sci Technol. 2014;57(1):188–93.
75. Barros F, Awika JM, Rooney LW. Interaction of Tannins and other Sorghum phenolic compounds with starch and effects on in vitro starch digestibility. J Agric Food Chem. 2012;60(46):11609–17.
76. Mkandawire NL, Kaufman RC, Bean SR, Weller CL, Jackson DS, Rose DJ. Effects of Sorghum (Sorghum bicolor (L.) Moench) Tannins on α-amylase activity and in vitro digestibility of sstarch in raw and processed flours. J Agric Food Chem. 2013;61(18):4448–54.

77. Mao S, Ren Y, Ye X, Kong X, Tian J. Regulating the physicochemical, structural characteristics and digestibility of potato starch by complexing with different phenolic acids. Int J Biol Macromol. 2023;253:127474.
78. Wang X, Liu L, Chen W, Jia R, Zheng B, Guo Z. Insights into impact of chlorogenic acid on multi-scale structure and digestive properties of lotus seed starch under autoclaving treatment. Int J Biol Macromol. 2024;278:134863.
79. Bello-Perez LA, Agama-Acevedo E, Garcia-Valle DE, Alvarez-Ramirez J. A multiscale kinetics model for the analysis of starch amylolysis. Int J Biol Macromol. 2019;122:405–9.
80. Meraz M, Alvarez-Ramirez J, Vernon-Carter EJ, Reyes I, Hernandez-Jaimes C, Martinez-Martinez F. A two competing substrates michaelis–menten kinetics scheme for the analysis of in vitro starch digestograms. Starch-Stärke. 2020;72(9–10):1900170.
81. Nguyen GT, Sopade PA. Modeling starch digestograms: computational characteristics of kinetic models for in vitro starch digestion in food research. Compr Rev Food Sci Food Saf. 2018;17(5):1422–45.
82. Butterworth PJ, Warren FJ, Grassby T, Patel H, Ellis PR. Analysis of starch amylolysis using plots for first-order kinetics. Carbohydr Polym. 2012;87(3):2189–97.
83. Vernon-Carter EJ, Meraz M, Bello-Perez LA, Alvarez-Ramirez J. Analysis of starch digestograms using monte carlo simulations. Carbohydr Polym. 2022;291:119589.
84. Blanquet S, Zeijdner E, Beyssac E, Meunier J, Denis S, Havenaar R, Alric M. A dynamic artificial gastrointestinal system for studying the behavior of orally administered drug dosage forms under various physiological conditions. Pharm Res. 2004;21:585–91.
85. Qin Y, Chen XD, Yu A, Xiao J. New understanding from intestinal absorption model: how physiological features influence mass transfer and absorption. AICHE J. 2023;69(8):e18099.
86. Vonk RJ, Hagedoorn RE, de Graaff R, Elzinga H, Tabak S, Yang YX, Stellaard F. Digestion of so-called resistant starch sources in the human small intestine. Am J Clin Nutr. 2000;72(2):432–8.
87. Muñoz-Tamayo R, Laroche B, Walter E, Doré J, Leclerc M. Mathematical modelling of carbohydrate degradation by human colonic microbiota. J Theor Biol. 2010;266(1):189–201.
88. Bucci V, Xavier JB. Towards predictive models of the human gut microbiome. J Mol Biol. 2014;426(23):3907–16.
89. Dona AC, Pages G, Gilbert RG, Kuchel PW. Digestion of starch: in vivo and in vitro kinetic models used to characterise oligosaccharide or glucose release. Carbohydr Polym. 2010;80(3):599–617.
90. Gouseti O, Bornhorst GM, Bakalis S. In: Mackie A, editor. Interdisciplinary approaches to food digestion. Springer; 2019.

Chapter 7
Starch in 3D Food Printing

Renan Nascimento Araújo, Samile Bezerra de Aguiar, Pedro Augusto Invernizzi Sponchiado, Sara Kierulff Balabram, Larissa Tessaro, Stanislau Bogusz Junior, and Bianca Chieregato Maniglia

7.1 Introduction

Three-dimensional printing, also referred to as additive manufacturing or Solid Freeform Fabrication (SFF), is a technique that is extensively utilized for the construction of three-dimensional objects in a layer-by-layer manner, with a high degree of precision and accuracy in all dimensions [1, 2]. The potential applications of this new technology are numerous and diverse, spanning a range of sectors including construction [3], aerospace [4], textiles [5], pharmaceuticals [6], automotive [7], packaging [8], medical [9] and food [10].

3D food printing represents an emerging technology that enables the creation of customized foods in terms of shape, texture, and nutritional composition [11]. In recent years, the use of edible materials such as starch has generated considerable interest in 3D food printing research due to their functional properties, including gelling and viscosity, which are essential for extrusion and the formation of complex shapes [12]. Starch, derived from botanical sources such as corn, potato, cassava, wheat, is a biopolymer that is widely used in the food industry [13]. Due to its controllable gelatinization, thickening, emulsifying, and stabilizing characteristics, starch has become a promising ingredient for the innovative side of additive manufacturing [14].

Native starches have distinct rheological properties that make them highly suitable for 3D printing of food products, but they also have a high hydrophilic character, limited pressure, heat, and shear resistance, as well as a tendency to retrograde

R. N. Araújo · S. B. de Aguiar · P. A. I. Sponchiado · S. K. Balabram · L. Tessaro
S. B. Junior · B. C. Maniglia (✉)
São Carlos Institute of Chemistry (IQSC), University of São Paulo – USP,
São Carlos, SP, Brazil
e-mail: biancamaniglia@usp.br

G. Ayala Valencia (ed.), *Starch: Progress in Food Applications*,
https://doi.org/10.1007/978-3-031-98732-8_7

after storage and possibly occur the syneresis [15]. However, various modification techniques can improve its properties to suit specific applications [16].

The functional properties of starch through physical, chemical, and genetic modifications with use of biotechnology have been working by researchers [17]. In recent years, there has been a notable increase in research activity focused on the modification of starches through the application of novel food processing technologies. The implementation of these techniques can result in the development of materials that are optimally suited for 3D printing applications. This may be achieved by enhancing their strength, flexibility, and thermal stability [18].

Furthermore, the incorporation of starch in 3D printing provides additional nutritional benefits. It is slowly digestible carbohydrate source that can be transformed into functional foods, such as those with a low glycemic index, and tailored to meet the specific dietary requirements of consumers with restrictions [19]. In this context, an important application of three-dimensional food printing is the production of food for individuals with dysphagia. Patients with dysphagia have difficulty swallowing their food safely, which increases the risk of choking and aspiration. These risks can result in malnutrition and pneumonia [20]. Therefore, the main strategy to mitigate these risks involves producing foods with a modified texture, which makes them safe for these individuals [20, 21]. Additionally, the additive manufacturing process enables the integration of other nutrients, including protein and fiber, which enhances the nutritional profile of printed foods [22].

In conclusion, the future of 3D starch food printing is contingent upon a more comprehensive understanding of the interactions between food ingredients and the technological requirements of the process. Starch represents a plentiful and appropriate raw material for 3D printing, and thus merits greater attention. This polysaccharide has two primary applications: it can be utilized as a direct printing material or as an additive to enhance the printing properties of other materials. Contemporary research focuses on three main areas when it comes to starch: the relationship between its structure and printability, repairs to a three-dimensional starch-based product, and its 3D printing capacity [17, 23].

7.2 Overview of 3D Printing Technology Using Starch

Additive manufacturing can be regarded as a digital production method that facilitates the creation of three-dimensional structures that can be customized in terms of composition, processing conditions, and geometry. The material's application, in accordance with the principle of layer-by-layer deposition, occurs directly onto the printing table from a model created by computer-aided design (CAD) software [24]. This technology integrates engineering and materials science with computer-aided design, a practice that involves the interdisciplinary application of various fields. The development of 3D printers capable of producing objects with the necessary precision and speed falls within the domain of engineering, while the virtual three-dimensional representations that include details relating to the positioning of the components of the final product are the responsibility of the CAD software. There are numerous modalities of 3D printing, including selective sintering, optical immobilization, extrusion, melt deposition, inkjet and binder injection. In the food industry, the most prevalent forms of 3D printing are extrusion, binder injection, and selective sintering [25]. Of these, extrusion is the most commonly utilized method.

7.2.1 Extrusion-based 3D Printing Technology

In the contemporary context of the food industry, extrusion 3D printing has emerged as the predominant 3D printing technology. The extrusion 3D printer's primary component is the nozzle, which contains a piston that propels the material through the nozzle. Subsequent to extrusion, the raw materials quickly curdle and adhere to the previous layer, thereby forming the finished product on the printer platform [26]. The nozzle, otherwise referred to as the platform, is capable of vertical movement in the x-y plane. This enables the creation of successive two-dimensional layers, which, through a layer-by-layer process, ultimately result in the formation of 3D objects. The extrusion-based additive fabrication process can be categorized into three distinct methods: screw, pneumatic, and syringe extrusion. These methods differ in the manner in which the ink material is propelled through the nozzle (Fig. 7.1). Syringe extrusion technology utilizes a motor that drives the piston, causing the material to pass through the nozzle. In contrast, screw extrusion involves the expulsion of ink from the die by rotating the screw. This method is particularly suitable for inks with low fluidity. Pneumatic extrusion, by contrast, forces the raw material through the nozzle by the pressure difference generated by the compression of the air. This method is particularly suitable for materials with high fluidity.

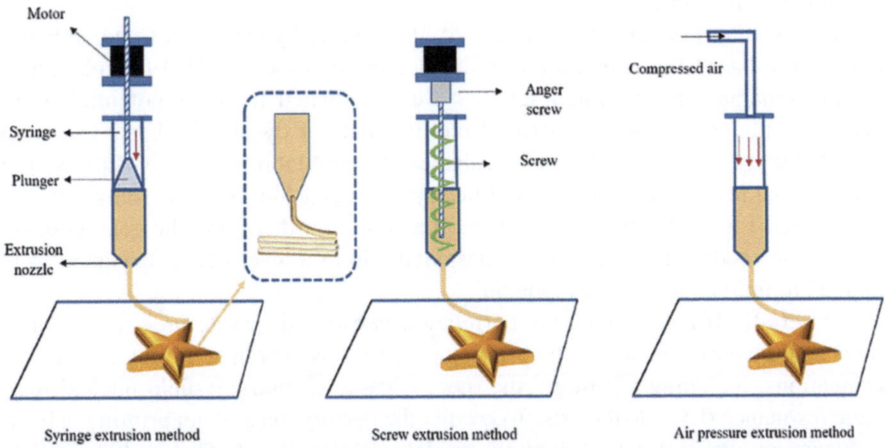

Fig. 7.1 Schematic diagram of the different types of 3D printing-extrusion: syringe, screw, and air pressure extrusion. (Reused with permission from Guo et al. [29])

Depending on the temperature employed during the process, extrusion 3D printing can be classified as hot-melt extrusion, room-temperature extrusion, or hydrogel-molding extrusion [27]. Hot melt extrusion is a technique frequently employed in the fabrication of inks that demonstrate reduced viscosity at elevated temperatures yet exhibit an increase in viscosity upon the reduction of thermal energy (e.g., certain fats). In this method, the nozzle and the container holding the material to be extruded are maintained at elevated temperatures, while the printing table is kept at room temperature or even lower [28]. The extrusion technique can also be applied at ambient temperature, provided that the ink possesses sufficient fluidity to be compressed from the die and maintain structural stability during printing. Inks exhibiting these characteristics are analogous to plastics, demonstrating solid behavior below the yield point but flowing above it. Extrusion 3D printing technology commonly utilizes hydrogels as a printing material, particularly biopolymer molecules, due to their capacity to gel through covalent cross-links.

Below, some works will be mentioned that involve the 3D printing process by extrusion involving starch and applications in the food area.

Maniglia et al. [30] investigated the use of dry heat treatment (DHT) to improve the 3D printing capacity of wheat starch hydrogels by extrusion. The starch granules were subjected to dry heat treatment at temperatures of 130 °C for intervals of 2 hours (DHT_2h) and 4 hours (DHT_4h). The employment of this eco-friendly technology resulted in substantial alterations in mechanical, rheological, and physicochemical characteristics of the hydrogels, indicating that DHT serves as a substantial modification process of enhancing the properties of wheat starch hydrogels, rending them suitable for utilization in 3D food printing applications.

Castillo et al. [31] applied extrusion 3D printing technology to create semi-solid foods made from pregelatinized starch with designed size, number of pores, and pore size. The use of starch leads to a drastic change in mechanical properties, which proved to be interesting for the food chewing process aiming for an application for specific consumers.

Guedes et al. [19] explored the use of 3D printing by extrusion to manufacture corn starch-based foods modified by dry heating treatment (DHT) for individuals with dysphagia. The modified starches were evaluated for their potential in 3D printing and for meeting the textural requirements of dysphagic diets. The study revealed that the DHT modification enhanced the printability of native corn starch. The optimal performance of the modified starch was achieved at the following heat treatments: 140 °C for 6 h, 150 °C for 4 h and 5 h. Furthermore, the gels produced from these modified starches were classified within a level that is expected to be applied to individuals with dysphagia.

Jiao et al. [2] investigated the correlation between disparate processing conditions and the performance of the printing process was appraised through a series of evaluations, including printing tests, rheological tests, and low-field nuclear magnetic resonance (LF-NMR) tests. To predict the performance of gel printing, a back-propagation artificial neural network model was developed. The study's findings offer both theoretical underpinnings for the utilization of corn starch gels in three-dimensional food printing and a novel methodology for predicting the printing performance of analogous materials. This methodology contributes to the optimization of printing parameters, thereby enhancing printing efficiency and quality.

In their studies, Zhong et al. [11] investigated the properties and 3D printing performance of starch gels with egg yolk powder. The researchers utilized egg yolk powder as the primary material, incorporating corn starch and potato starch at varying concentrations (10%, 15%, and 20%, respectively). The findings indicated that the composite material comprising the highest percentage of starch from diverse botanical sources exhibited the highest storage modulus (G′), loss modulus (G″), hardness, and apparent viscosity. The study found that egg yolk powder gels with corn starch (EYP-CS) at varying concentrations exhibited optimal swallowing performance, thereby meeting the criteria for classification as level 5 ingested foods.

Xu et al. [32] investigated the impact of freezing on the 3D printing performance of cassava starch gels, focusing on the multiscale structure and properties of the gel. When freezing occurred prior to the subcooling period, changes in 3D printing accuracy were primarily influenced by starch retrogradation. During this phase, the crystallinity, short-range ordered structure, structural recovery, and storage modulus of the starch were enhanced by increasing the proportion of short and medium amylose chains (X~100–5000). Consequently, the enhanced rigidity and structural stability of the gel led to superior printability of the material. However, the expansion of ice crystals during and after the freezing time, in the subcooling period, disrupted the short-range ordered structure. This disruption led to an increase in pore size, greater structural laxity in the gel microstructure, an increase in free water content (A23), and a drop in storage modulus. Consequently, the water retention capacity and structural stability of the starch gel were diminished, thereby reducing its capacity for 3D printing. This study offers valuable insights into the process of freezing starchy foods for 3D printing applications.

7.2.2 Selective Laser Sintering Printing Technology

Three-dimensional printing by sintering is a manufacturing method that involves the compacting and heating of particles of a powdered material to form a solid part (Fig. 7.2). In a closed chamber, the 3D object is created by a layer-by-layer deposition process on the powder bed using a laser [33]. Typically, post-processing is necessary to enhance the mechanical properties (e.g., tensile strength and hardness) or visual appearance (e.g., surface finish and dimensional accuracy) of the printed part. Such post-processing may include surface treatments, polishing or coating [34]. However, concerns have been raised about potential food safety issues with laser sintering, as the laser may alter the physical and chemical properties of food materials.

Jonkers et al. [35] investigated the use of laser sintering in the 3D printing of starch-based foods, focusing on the structural and mechanical properties of the printed products. The mechanical behavior and microstructure of these foods were analyzed using uniaxial compression tests and computed tomography, considering

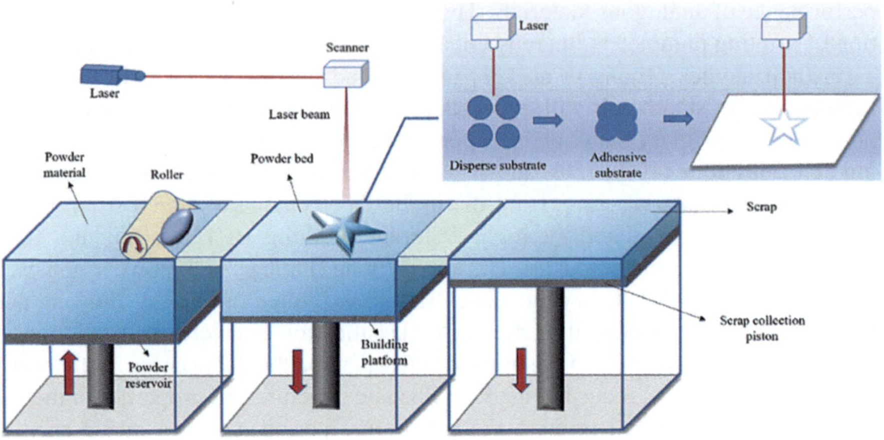

Fig. 7.2 Schematic diagram of selective laser sintering 3D printer. (Reused with permission from Awad et al. [34])

two orientations: the build direction and the in-plane principal direction. By increasing the areal energy density from 14 J/cm^2 to 82 J/cm^2, the build direction stiffness more than tripled, while the in-plane stiffness remained constant, and the fracture stress more than doubled. However, this also resulted in reduced ductility in the build direction. These effects were correlated with changes in the microstructure, such as an increase in the average relative density from 42% to 50%, and variations in local mechanical properties. Furthermore, heterogeneous crack propagation was observed in the material during *in situ* compression tests.

7.2.3 Binder Jetting Printing Technology

In this 3D printing technology, a binding liquid is applied to a bed of powdered material in specific regions, causing the particles to blind together (Fig. 7.3). The 3D material is constructed through the successive application of food-grade powder, which is deposited in a layering technique. The counter-rotating drums facilitate the uniform distribution of powdered material across the printing surface, enabling the jet printing head to disperse the binder onto predefined regions, as delineated by CAD software. This process leads to the agglomeration of adjacent particles, resulting in the formation of a 2D pattern on the layer. After curing, the powder bed and support platform are lowered to accommodate the next layer of powder, which is deposited on top of the previous layer [36, 37]. This technique provides a low-coast and rapid method for creating objects, but it has notable limitations, including poor resolution, a rough surface finish, and high equipment costs [26]. In addition, binder jet printing is limited to powdered materials [38]. The quality of the final 3D printed

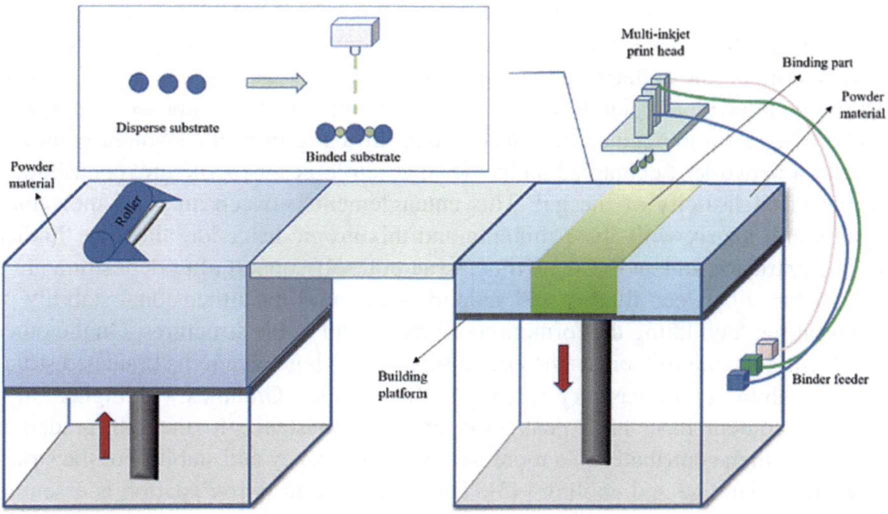

Fig. 7.3 Schematic diagram of binder injection 3D printer. (Reused with permission from Jiang et al. [40])

object is directly influenced by the properties of the powder, including but not limited to particle size, bulk density, shape, fluidity, and surface properties [39].

Zhu et al. [41] explored the use of this 3D printing technology to produce foods rich in calcium caseinate powder. The researchers achieved adequate printability with a composite material consisting of calcium caseinate, medium-chain triglycerides, and starch powder. The printability of the blends was improved by the addition of starch, which reduced the swelling during the addition of the binder. Through the strategic modification of the composition of the binder, the proportion of ingredients, and the post-heating treatment, protein-rich food models were successfully produced, exhibiting a range of textures from crumbly to elastic. This research underscores the potential for the development of high-protein foods through the utilization of binder jet 3D printing technology.

7.3 Structural Properties of Starch for 3D Printing

7.3.1 Amylose and Amylopectin Impact

The behavior of starch during 3D printing processes is largely influenced by its two main fractions: amylose, an essentially linear polysaccharide, and amylopectin, a highly branched polysaccharide. Amylopectin is composed of D-glucopyranose monomers linked by α-(1,4) or α-(1,6) glycosidic linkages and make up 70–85% of conventional starch. While amylose, a linear chain of α-D-glucopyranosyl units linked by (1→4) bonds, is present in smaller proportions, around 15–30% of common starch, playing an important structural role in starch [42]. In extrusion 3D printing, the behavior of starch, especially its amylose and amylopectin fraction, plays a critical role in defining the printability and rheological properties of the ink. During gelatinization, starch forms a continuous gel matrix composed of amylose and amylopectin molecules. Amylose, when diffusing from the swollen granules, creates a network of entangled molecules that reinforces the structure, contributing to the viscoelasticity of the gel. This entanglement between amylose molecules results in a matrix with shear thinning and thixotropic behavior, allowing fluidity during extrusion and, at the same time, structural self-support after deposition [17]. This balance between fluidity and support is essential for dimensional stability in 3D printing, facilitating the formation of precise and stable structures. On the other hand, amylopectin influences the printability of starch gel due to its branched structure and ability to increase crystallinity in the granules. Granules with higher amylopectin content have high peak viscosity and resistance to the retrogradation process, which contributes to a more uniform consistency and stability of the structure during storage and cooling [43]. This resistance to retrogradation is essential for maintaining the structural properties of the gel after printing, reducing deformations that could compromise the accuracy of the printed parts.

In addition, controlling the starch concentration is essential to achieve ideal extrudability and stability properties in 3D printing, and these characteristics vary significantly depending on the amylose and amylopectin content in the starch. Starches with low or no amylose content produce stickier textures and stable, translucent pastes, which are ideal for applications that require high adhesion and uniform consistency, such as in papermaking. On the other hand, starches rich in amylose, characterized by greater resistance to digestion and film-forming properties, offer a firm texture and are less susceptible to shear thinning [44].

7.3.2 Molecular Structure Impact

Amylose, with its linear structure, has structural properties that influence the formation of ordered structures, such as chain length and molecular weight of amylose: short chains form more rigid gels due to increased entanglement between chains, while longer chains promote broader crystalline structures. In aqueous solution, its helical conformation allows the formation of supramolecular complexes and entangled networks, which increase starch viscosity [45]. These properties influence not only the consistency of starch, but also its behavior during processing, being especially valuable in applications that require texture control and structural strength, such as 3D printing. Furthermore, in the presence of other molecules, such as lipids, amylose forms complexes that limit leaching during heating, reducing granule swelling and contributing to the maintenance of the structure. At gelatinization temperatures, the presence of amylose reduces the melting point, and the amount of energy required, facilitating the formation of pastes and promoting a more controlled phase transition [46].

Amylopectin, on the other hand, is a branched molecule that increases the crystallinity of starch when present in a greater proportion in granules, resulting in more compact structures that are less prone to retrogradation, especially during refrigerated storage. Granules rich in amylopectin, such as waxy starches, have high peak viscosity and clarity in pastes, properties enhanced by the presence of monoester phosphates [46]. The presence of phospholipids in amylopectin makes pastes opaque and reduces their viscosity, an important effect for formulations that require specific transparency and thickness characteristics [47].

The structure and size of the starch chains (molecular weight) directly influence the viscosity of the solution and therefore the processability of the 3D printing equipment. Longer starch chains, with higher molecular weight, tend to increase the viscosity of the solution, which can hinder the flow of the gel, compromising the printing accuracy. On the other hand, shorter chains result in a lower viscosity, facilitating the extrusion process and allowing greater control during layer formation [48]. Therefore, modifying the rheological properties, through adjustments in the size of the starch chains and the distribution of the amylopectin and amylose chains, directly affects the viscosity and gelation behavior, influencing the rheology of the starch solution. This, in turn, has a direct impact on the ability of the material to flow properly during the 3D printing process, on the structural stability and definition of the printed products [17].

Therefore, the behavior of the amylopectin and amylose fractions in starch is essential for the success of extrusion 3D printing, as each one plays a specific role in the rheological and structural properties of the starch ink. Amylose contributes an entanglement network that confers viscoelasticity and dimensional stability, while amylopectin provides crystallinity and resistance to retrogradation, critical elements for maintaining the shape and stability of printed parts. Precise control of concentration and temperature during the printing process optimizes these effects, resulting in highly accurate and self-supporting structures. Thus, the interaction between amylose and amylopectin enables the creation of highly efficient starch gels, highlighting starch as a promising material for 3D printing applications.

7.4 Methods to Improve the Processability of Starch in 3D Printing

7.4.1 Technologies for the Modification of Starch

In the field of 3D printing, the thermal, rheological, and surface properties of starch can impact directly the processability and the quality of final product [49]. In this context, natural starch exhibits remarkable rheological properties that make it suitable to produce edible inks for 3D printing foods [49]. Starch from different botanical sources can present distinct characteristics, especially regarding their rheological properties. Corn and sweet potato hydrogels, for example, exhibits pseudoplastic behavior, which allows these materials to flow through a needle when pressure is applied to the system and solidify once the pressure is over [50].

However, studies have highlighted certain properties of starch as ideal for using this polysaccharide as a raw material for inks, including the adjusted swelling behavior of its hydrogels, low gelatinization temperature, strong gels formation, and adequate pasting properties [23, 51]. Intending to improve some of the limited properties of starch that are important factors for 3D printing technology, such as low thermal, shear endurance and pressure, as well as high propensity to retrograde after storage, some modification treatments can be applied, such as physical, chemical, enzymatic, and even genetic treatments, as summarized in the Table 7.1.

Physical methods of modification can act by altering from the crystalline structure of the starch granules to some physicochemical properties [61]. These treatments include thermal, such as dry heating and heat moisture treatments, and non-thermal treatments, like high pressure processing and pulsed electric field.

Dry heating treatment (DHT) is a simple and relatively cheap method that is considered safe and "green", since it is a pollution-free process that does not use chemical reagents and does not generate waste [62]. This physical approach involves low moisture content (normally between 5% and 12%) and heating the samples to high temperatures (110 °C to 150 °C) for durations that can range from minutes to hours. Several studies have investigated the effects of DHT on the physicochemical properties of different types of starch, such as cassava [63], potato [64], quinoa [65], and blue highland barley [66]. According to the authors, this method can significantly affect the relative crystallinity, solubility, swelling power, gelatinization properties, and pasting viscosity of starch. Guedes et al. [19] studied the effects of different conditions of DHT (110–150 °C, between 1 and 6 h) over the 3D printing, mechanical and physicochemical properties of maize starch with the aim of use the modified materials as gelling ingredients for fruit-based foods for individuals with dysphagia. At the optimized conditions (140 °C for 6 h, and 150 °C for 4 h and 5 h), the authors obtained modified materials with new interactions among starch molecules, with better printability and improved paste properties.

Like the DHT, that is considered safe and an environmentally friendly process, the heat-moisture treatment (HMT) can be also applied to make changes in the physicochemical properties of starch. As a hydrothermal technique, the HMT is used with higher moisture content than DHT (12–35%), and temperatures between 80 °C and 140 °C for certain durations [67]. Zheng et al. [68] observed an improvement in the accuracy of 3D printing HMT modified wheat starch (120 °C for 2 h, with moisture content adjusted to 30%), in the height of printed samples, and prolonged the storage time of the developed materials with softer texture.

Differently of the DHT and HMT, for example, the high hydrostatic pressure (HHP) or high-pressure processing (HPP) is a non-thermal process that is mainly used for microbial inactivation in different types of food in industry [55]. However, for starch, the high-pressure environment (100–600 MPa, for 2–30 min at room temperature) can modify the functional properties in such a way that it may alter the pasting properties and rheology of starch hydrogels, making it more suitable to be used as raw materials for 3D printing food [23]. Larrea-Wachtendorff et al. [55]

Table 7.1 Summary of different types of modifications used to improve the properties of starch

Type of starch	Treatment	Conditions used in the treatment	Improvements	Reference
Wheat	Physical—Dry heating (DHT)	130 °C, 2 h and 4 h	Lower apparent viscosity of pastes; Greater structural strength; Reduced syneresis; Improved printability	[52]
Wheat	Physical—Heat moisture (HMT), ultrasonication (UT) and freezing (FUT)	HMT: 90 °C, 2 h and 4 h UT: 20 °C, 10 min and 20 min, 50% of pulse (10 s ON, 2 s OFF) FUT: –18 °C for 12 h, followed for ultrasonication (750 W, until reach a temperature of 25–30 °C)	HMT promoted an increase in swelling power and a decrease in synerisis of materials; UT promoted depolymerization and reduction in cristallinity; FUT improved the pasting properties of materials and resulted in the best printability	[53]
Wheat and cassava	Physical—Pulsed electric field (PEF)	Pulses at: 15 kV/cm and 25 kJ/kg; 25 kV/cm and 25 kJ/kg; 25 kV/cm and 50 kJ/kg (optimized condition), 5 μs pulse length	Wheat: reduction of peak apparent viscosity; firmer gels; reduction of gelatinization enthalpy Cassava: reduction of peak apparent viscosity; reduction of gelatinization enthalpy	[54]
Potato	Physical—High hydrostatic pressure (HHP) combined with thermal treatment	HHP: 600 MPa for 15 min, at 25 °C; Thermal treatment: 40 °C and 50 °C	Stronger gels obtained under HHP-thermal treatment; increase in the viscosity; improvements in rheological properties (G′>>G″)	[55]
Cassava	Chemical—Ozonation	43 mg O_3/L, flow rate 1 L/min, 15 and 30 min	At optimized conditions (30 min), the gels of ozonated starch showed lower pH, lower peak apparent viscosity, stronger and more printable gels (at high % starch)	[14]
Barnyard millet	Chemical—Crosslinking	Sodium trimetaphosphate (STMP) used in three conditions (1, 3 and 5% w/w)	Lower swelling power and solubility; increased in gelatinization temperature, but decreased the enthalpy of gelatinization; films with lower water vapor permeability and higher tensile strength	[56]
Corn	Chemical—Esterification	Synergistic crosslinking Fe^{+3}-Citric acid, at different concentrations (CA:Fe^{+3} at 1:0.2, 1:0.4, 1:0.6, 1:	At the optimized condition (1:0.6): increased on elastic modulus (G′), dense honeycomb porous structure, small pore size distribution	[57]

(continued)

Table 7.1 (continued)

Type of starch	Treatment	Conditions used in the treatment	Improvements	Reference
Rice	Enzymatic	4-α-Gtase for 1, 24, and 96 h	Alterations in gelation properties; decreased viscosity	[58]
Oat	Enzymatic	*Bacillus acidopullulyticus* pullulanase	Reduced digestibility, decreased viscoelasticity	[59]
Potato	Genetic	*GBBS* gene (granule-bound starch synthase gene), *SBEI* and *SBEII* (starch branching enzyme)	Improved functionality of modified starches, with higher amylose content	[60]

submitted potato starch suspensions (20% w/w) under HHP at 600 MPa, for 15 min and 50 °C, and obtained a material with complete gelatinization, gel-like appearance, high viscosity and a highly structured profile (G′>>G″) without the need of high temperatures.

In addition to physical methods, chemical treatments can also be applied to achieve the desired functional properties of the materials. As a polysaccharide formed by glucose units, starch is rich in hydroxyl groups that can be replaced with different functional groups at position C2, C3 and C6, creating new intermolecular interactions in the starch structure [69]. Intending to improve the functional properties of starch through chemical treatments, the most used methods usually involve the oxidation through ozone technology, for example, crosslinking and esterification [70].

Considered an environmentally friendly technology, the ozone process is a powerful tool to oxidizing starch, whether reacting directly (between the substrate and molecular O_3), or indirectly (normally in alkaline medium), leaving no toxic residues [71]. Lima et al. [72] submitted cassava starch in the ozonation process (43.5 mg $O_3.L^{-1}$, for 15, 30 and 60 min) and observed the formation of carbonyl and carboxyl groups, as well as physicochemical changes such as a reduction of the viscosity, reduction of the starch retrogradation and around 280% stronger hydrogels than the native starch hydrogels.

Crosslinking in starch increases the degree of polymerization and molecular mass through the interconnection of linear and branched chains. Several agents can be used, such as sodium trimetaphosphate (STMP), phosphoryl chloride ($POCl_3$), and some poly-functional carboxylic acids as citric acid [70, 73]. Nakkala et al. [74] evaluated the effect of different chemical reagents (oxidation with sodium hypochlorite, succinylation with sodium octenyl succinic anhydride, and phosphorylation with sodium tripolyphosphate—STPP, and sodium trimetaphosphate—STMP) on the physicochemical properties of turmeric starch. The results showed that all chemically modified starches had lower percentage of syneresis than the native starch, and the esterification/crosslinking modifications improved the paste properties, leading to an increase in viscosity, especially for the starches treated with STMP.

In addition to physical and chemical treatments, it is possible to employ enzymatic and even genetic processes to modify starch properties. Enzymatic processes, also considered as environment friendly methods, can modify the starch structure through the break of glycosidic bonds and the formation of new intermolecular interactions [23]. For reach this purpose, there is a several of enzymes that can be used, such as endo-amylases (α-amylase), exo-amylases (β-amylase), and debranching enzymes (pullulanase) [75]. Shah et al. [76] obtained a decrease in crystallinity and a shear-thinning behavior of dual enzyme modified (β-amylase and transglucosidase) of three oat starches, which is an important rheological factor for 3D printing. Geng et al. [77] used two-step modification on rice starch, with pullulanase (PUL) and transglucosidase (TG), and obtained a compact gel, with improved gel strength and reduced digestibility.

7.4.2 Additive in the Starch-formulation for 3D Printing

The incorporation of additives has been demonstrated to enhance the properties of starch gels, thereby serving as a promising alternative to conventional chemical or physical modification processes of this polysaccharide [78]. These additives can substantially modify the structural characteristics of starch, enhancing its performance in 3D printing applications. Polysaccharides of animal or plant origin are bioactive natural products with high molecular masses, and they are being widely exploited for the development of healthy foods due to their interesting biological properties [79]. The microstructure of native starch and its gel properties can be enhanced by the incorporation of these polysaccharides [80]. The structural stability of the printed form is enhanced due to the formation of more compact three-dimensional network structures in the starch-polysaccharide matrix, resulting in an increase in the storage modulus (G′) and, consequently, an increase in the mechanical strength of the material [81]. The viscosity of starch-polysaccharide composite gels is generally increased due to the thickening properties of the polysaccharides. However, this heightened viscosity can impede the extrusion of starch inks into the printing matrix [82]. Cui et al. [82] found that the addition of xanthan gum and sodium alginate (2.5%/2.5% w/w, dry basis) improved the texture, viscosity, and loss modulus (G″) of potato starch gels, while maintaining the structural stability of the printed form and denoting optimum geometric precision. In summary, Chen et al. [83] proposed that polysaccharides are effective additives in improving the sensory quality, nutritional value, rheological properties, and thermal stability of materials for 3D food printing. The advancement of food 3D printing has been constrained by the materials employed as inks, which is attributable to the complexity of utilizing these materials directly for printing or with constrained manipulation capabilities. In this context, polysaccharide gels offer enhanced fluidity and substantial improvements in printing properties, irrespective of their role as a primary or secondary material. Consequently, these composites have the capacity to influence starch digestibility, thereby decreasing the content of rapidly digestible starch

(RDS) and increasing the content of slowly digestible starch (SDS) and resistant starch (RS). This property is of particular relevance to starch-based food products, as it is associated with metabolic diseases such as diabetes and obesity [84]. It is noteworthy that the positive effects of polysaccharides on the digestibility and structure of printed foods are attributed to the following factors: Firstly, starch rearrangement is favored during the paste formation process, increasing the content of SDS and RS, which are less susceptible to degradation by amylase. Secondly, the increased viscosity of starch gels affects sensory attributes, palatability, and satiety [85]. Thirdly, α-amylase activity can be inhibited by the addition of these polysaccharides through competitive and non-competitive inhibition mechanisms [86]. Consequently, the incorporation of these polysaccharides into starch profoundly impacts the quality of products fabricated through three-dimensional printing [81]. Given the enhanced nutritional benefits of these additives and the customizable nature of 3D printing, future research could focus on developing products with adjustable digestibility and tailored to the specific needs of the target audience.

In addition, salts can be regarded as additives in the formulation of starch-based inks for three-dimensional printing. These salts are essential minerals that are required by the human body and primarily affect the structural properties of starch. Their incorporation has been shown to reduce water activity and the formation of hydrogel bonds, thereby significantly altering the paste properties and gelatinization temperature of starch [87]. The use of potassium chloride (KCl) and sodium chloride (NaCl) has been demonstrated to enhance the properties of maize starch-MCP gels, underscoring the efficacy of salt ions in improving printability. Furthermore, calcium chloride ($CaCl_2$) has been observed to promote cross-linking between the Ca^{2+} ions and the amide groups in the gel under microwave heating, leading to aggregation of the gel. This, in turn, results in an increase in its plasticity, a reduction in viscosity, and facilitates extrusion and stabilization of the shape during printing [88]. Lipids, which are prevalent in natural and processed foods, play a pivotal role in determining the palatability of foods [83]. Moreover, they have been utilized as additives to enhance the printability of bio-based inks, thereby rendering the starch paste softer and more amenable to extrusion, while concomitantly reducing the hardness and toughness of the printed mass [80]. Glycerol has been shown to contribute to a more continuous network structure by disrupting inter- and intra-network connections between polysaccharide chains [89]. The liquidity of the printing material can be enhanced by the segregation and formation of larger fat droplets under shear forces, which are subsequently distributed on the surface during printing. The shape stability and printing precision of semi-skimmed milk are superior to those of skimmed milk powder. This is explained by the action of the fat present in semi-skimmed milk, which acts as a plasticizer or lubricant, facilitating a smooth flow of the material on the 3D printing needle [90].

Natural food proteins are frequently utilized in the fabrication of starch gels, a practice that is substantiated by their exceptional gelling properties. Proteins are indispensable in the food processing industry and play a pivotal role in determining the consistency of gel-based foods [91, 92]. They function as pivotal constituents in 3D printing inks due to their distinctive structural characteristics, in addition to their substantial biomedical and clinical values [93]. To a lesser extent, proteins form covalent bonds (e.g., Maillard reactions) and non-covalent interactions (e.g., hydrogen bonds, hydrophobic interactions, and electrostatic forces) that allow them to encapsulate the starch or adsorb to its surface. Consequently, the gel properties of the starch-protein system are enhanced, while digestion is impeded [87, 94]. The 3D printing capacity of biological materials has been enhanced by the incorporation of proteins. For instance, it has been observed that the utilization of pea protein isolate leads to a substantial increase in the entanglement between the banana matrix and the protein, resulting in enhanced dynamic modulus, adhesion strength, and reduced recovery and stress at break during ink performance [95]. The combination of whey protein isolates and gelatine yields printed yogurt gels that are more stable and more precisely shaped [96]. Hussain et al. [97] found that the addition of pea protein isolate facilitated the flow of starch-based materials through the nozzle, thereby enhancing their structural stability after deposition. As previously mentioned, proteins have been shown to positively influence the structure and functional properties of starch, thereby enhancing its printability. Proteins are essential macronutrients and integral components of the human body. Their incorporation is beneficial to protein intake. Chuanxing et al. [98] have suggested a positive correlation between proteins and the degree of cross-linking, cohesion, adhesiveness, and thermal properties of food 3D printing materials.

7.5 Optimization of 3D Printing Parameters

7.5.1 Concentration of the Inks

The concentration of inks is a crucial factor to be evaluated in the context of printability, particularly in the case of Direct Ink Writing (DIW) 3D printers [99]. The fundamental operational principle of these printers' hinges on the extrusion of the material to be deposited on the printing surface. Consequently, the inks utilized in such printers must possess adequate plasticity to be extruded through the nozzle [100]. The concentration of the component of interest present in food inks, such as starch, has a direct impact on the rheological properties and, consequently, the extrudability of the material [101]. Park et al. [102] investigated a novel methodology for utilizing puffed rice as an innovative and sustainable ink for three-dimensional food printing. The food inks were evaluated for printability at the following concentrations of puffed rice: 15, 20, 25, 30 and 40%, as illustrated in Fig. 7.4. The rheological analysis of the food ink with 25% puffed rice revealed an optimal viscosity (η = 897.4 Pa.s), flux stress (τ_f = 1509.2 Pa) and yield stress (τ_y = 2471.3 Pa), demonstrating superior viscoelastic properties that facilitated printability and prevented the layers from collapsing, thus ensuring good shape stabilization after printing.

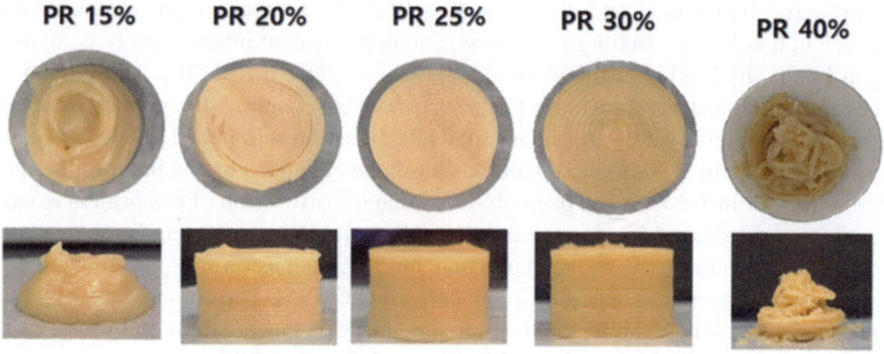

Fig. 7.4 Structures of extrusion-based 3D printing by PR paste (15, 20, 25, 30 and 40%), upper and side view (2 cm × 2 cm). (Reused with permission from Park et al. [102])

7.5.2 Nozzle Size

The quality of constructions printed with starch ink is directly affected by nozzle size. Consequently, variations in nozzle size design have a significant impact on printing performance [103]. The diameter of the nozzle affects several parameters, including size, shape, visual appearance, surface roughness, and printing precision [80].

In instances where the nozzle diameter is inadequate, ink accumulation at the tip is observed due to insufficient compression pressure. Conversely, when the nozzle diameter is excessively large, more ink than anticipated is ejected. Accordingly, an investigation into the optimal diameter is warranted to ensure that the printed material exhibits a uniform geometry and a well-distributed internal filling texture [80].

The researchers Bao et al. [104] developed smart packing for the visualization of food freshness through the 3D printing of type Direct Ink Writing (DIW). The incorporation of blueberry anthocyanin-phycocyanin (BA-PC) as an indicator and chondroitin sulfate (CS) as a co-pigment/co-carrier in a potato starch (PS) matrix resulted in a favorable response when a 3D-printed smart colorimetric film (CF) was used. The colorimetric film was optimized by modulating the size of the printing nozzle from 0.2 mm to 0.6 mm, as illustrated in Fig. 7.5. The nozzle with a diameter of 0.4 mm demonstrated a printed colorimetric sensor with a more homogenous and smooth gel structure, as well as enhanced sensitivity to the response of ammonia, the molecule responsible for analyzing the freshness of salmon and beef. Considering these findings, it can be concluded that the applications of 3D printing in the field of smart food packaging have been expanded.

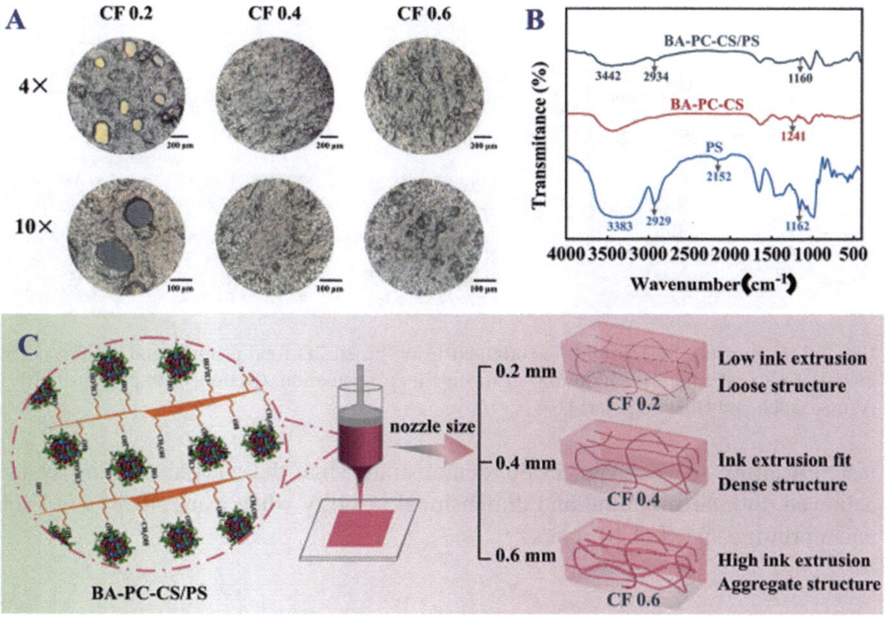

Fig. 7.5 Micromorphology and structural characterization of colorimetric film. (**a**) Image observations of CF 0.2, CF 0.4, and CF 0.6 under 4 × and 10 × objectives optical microscope; (**b**) FT—IR spectra of PS, BA-PC-CS and BA-PC-CS/PS; (**c**) Schematic diagram of colorimetric film component interactions and formation of different gel structures. (Reused with permission from Ma et al. [105])

7.5.3 Printing Speed

Printing speed represents another crucial parameter that must be subjected to meticulous analysis when optimizing 3D printing parameters. This is because it is directly correlated with the movement of the robotic arm across the three Cartesian coordinate a-axes [25]. This parameter determines the precision of the size of the 3D printing component [17]. Among the range of applications for starch-based food inks, printing speeds ranging from 6 to 60 mm/s are commonly used [106].

The emergence of food customization trends has prompted the development of novel food products with the objective of enhancing health through the integration of bioactive compounds (BCs). These compounds offer a spectrum of health benefits due to their antioxidant, anticancer, anti-inflammatory, antimicrobial, and immunomodulatory properties [107]. In this context, researchers [108] evaluated the optimization of printing speed by encapsulating lutein in double-layer starch/zein gels using a coaxial extrusion-based 3D food printing technique, as illustrated in Fig. 7.6.

The 3D printing process was evaluated at various speeds (4, 8, 14, and 20 mm/s) and zein concentrations (20%, 40%, and 60% w/w). Figure 7.7 presents images of 3D prints composed of starch (10%) and zein (20%, 40%, and 60%) at the four

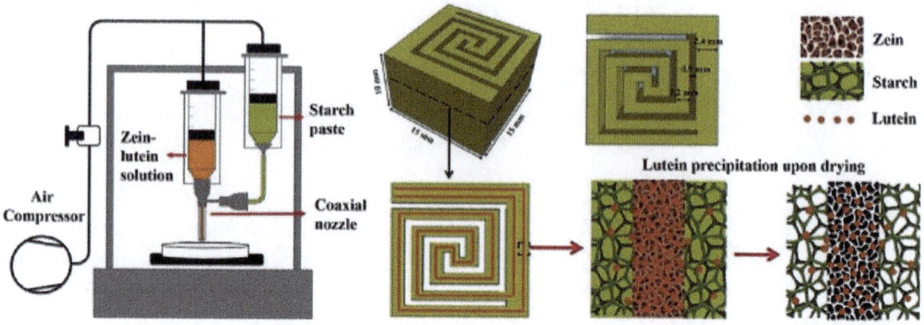

Fig. 7.6 Schematic illustration of coaxial extrusion-based 3D food printing and a 3D-rendered model of a spiral cube design optimized for lutein encapsulation. (Reused with permission from Ahmadzadeh and Ubeyitogullari [108])

tested printing speeds. The results demonstrated that the coaxial printing system achieved optimal structural and dimensional stability when using 40% (w/w) zein and a printing speed of 14 mm/s.

7.5.4 Flow Rate

The flow rate represents a key optimization parameter, closely linked to the distribution of filament thickness and affecting the distribution of pores in the print product [106]. For a given printing speed, an increase in the flow rate results in a reduction in porosity and a corresponding increase in the dimensions and total volume of the printed product [80].

A team of researchers [109] conducted an evaluation of the speed and flow parameters of 3D printing for the development of nutritious fruit-based snacks designed to provide 5–10% of the energy, calcium, iron, and vitamin D needed by children aged 3–10. The print flow was evaluated at 70, 100, and 130%, and the results demonstrated that the structural uniformity of the snacks improved at higher flow levels, exhibiting full interconnectivity between the cross filaments, as shown in Fig. 7.8, which represents cross-section X-ray images of 3D printed snacks for a more objective comparison.

7.5.5 Filling Rate

In order for a material to be successfully 3D printed, it essential that it is supported by an internal structure that is capable of maintaining its stability. This necessitates the use of a pattern and a fill rate [17]. Accordingly, the fill rate exerts an influence on the configuration of the ultimate 3D-printed item. Xiao-huan Chen et al. [111]

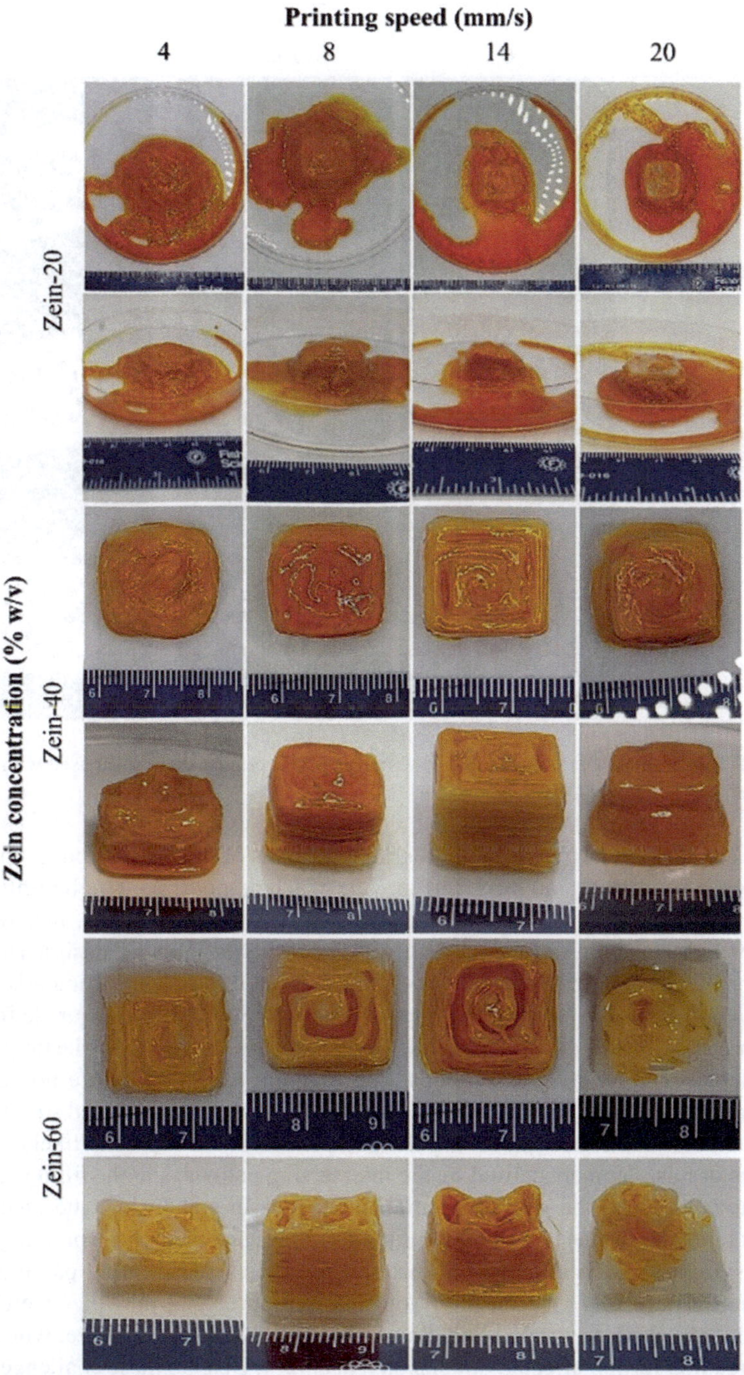

Fig. 7.7 Images of 3D prints made with starch (10%), and zein (Z; 20, 40, and 60%) at four printing speeds (4, 8, 14, and 20 mm/s). (Reused with permission from Ahmadzadeh and Ubeyitogullari [108])

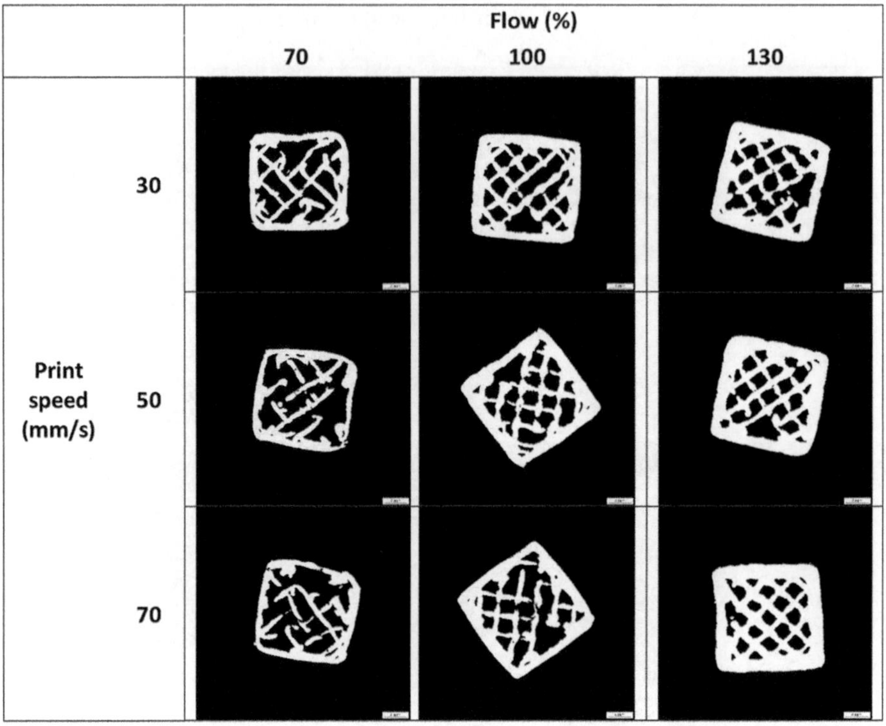

Fig. 7.8 Representative cross-section X-ray images of 3D printed snacks. (Reused with permission from Derossi et al. [110])

makes the comparison between photographic images of the 3D-printed samples and their corresponding digital models (Fig. 7.9) which demonstrated high printing precision. The overall appearance and structure of the printed samples, including the number and size of interior voids, closely aligned with the intended design. However, additional extruded lines were observed in the printed samples, particularly in grid and triangular patterns with filling rates of 25% and 50%. This issue arose from the continuous extrusion of ink during the 3DFP process. In polymer plastic printing, such defects can be mitigated by enabling filament retraction before non-printing movements [112]. Additionally, slight twists in the originally designed straight lines were evident in the grid and triangular samples. These distortions likely resulted from layer misalignment caused by the intersecting pathways in the filling sections. Such issues reduce the overall printing precision. In conclusion, the analysis of dimensional deviations between 3D-printed samples and their corresponding digital models highlights the influence of material properties and printing parameters on precision. The strong viscoelasticity of pumpkin puree and the "die-swell" effect contributed to larger extruded line dimensions than the set nozzle size, while material spreading further affected dimensional accuracy. Despite these challenges, most deviation values remained below 7%, indicating high overall printing precision.

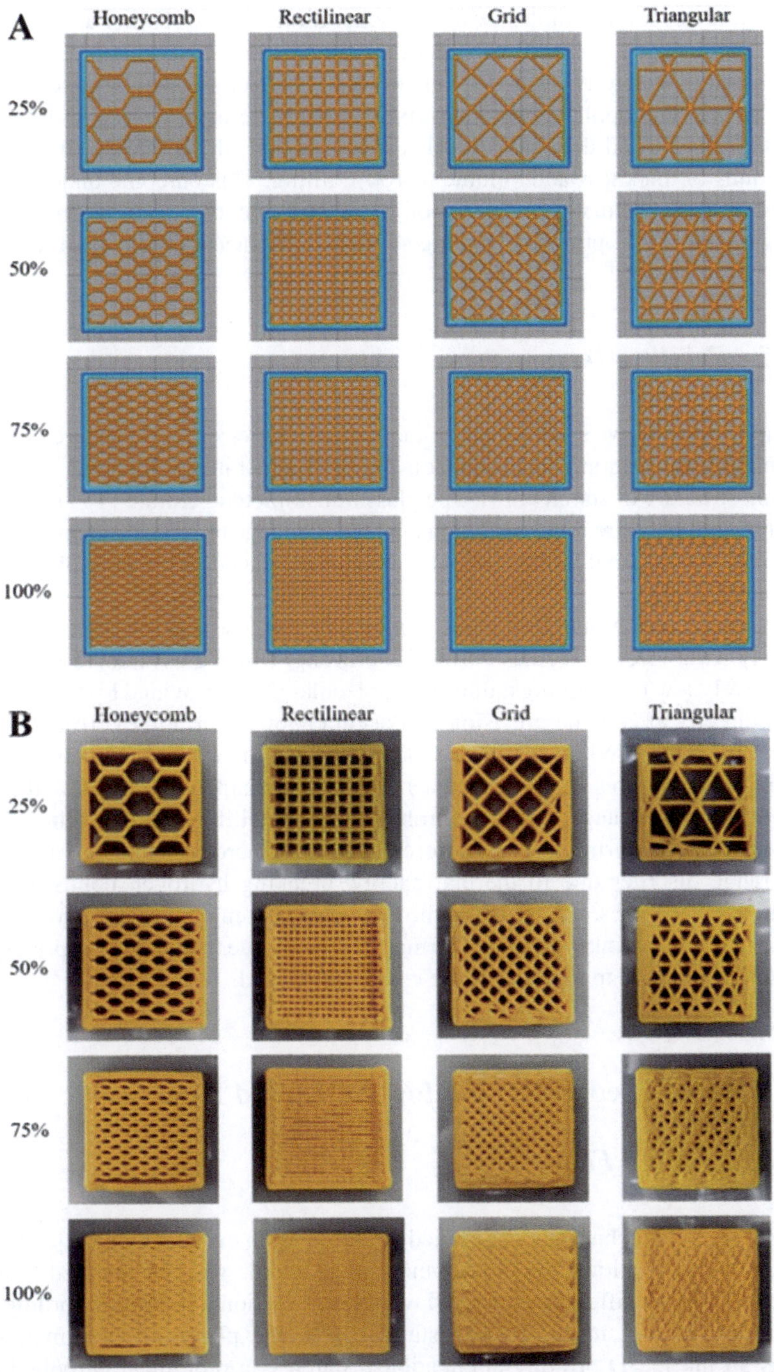

Fig. 7.9 The digital models (**a**) and the corresponding printed samples (**b**), showcasing various filling patterns and filling densities. (Reused with permission from Chen et al. [111])

Notably, honeycomb samples exhibited better dimensional stability at higher filling rates, with reduced length deviations as the filling rate increased from 25% to 75%. However, the structural instability due to weak connections between the filling and shell limited their applicability. Conversely, other patterns, such as grid and triangular samples, showed decreased precision with higher filling rates, with significant deviations in thickness and volume at 100% filling. This outcome underscores the trade-offs between material deposition, structural integrity, and precision, emphasizing the need for optimized printing settings to balance these factors effectively.

7.5.6 Printing Temperature

Multiple studies have shown that the thermal properties of food inks are crucial for optimizing 3D printing hot extrusion [113]. Temperature inherently affects the internal structure of starch [114]. The glass transition temperature (Tg) is an indicator of the temperature range in which a material changes from a rigid and hard state to a more flexible and malleable state and is thus related to the amorphous part of the matter [115].

In case of starch, its constituent granules of amylose and amylopectin interact strongly with water molecules during 3D printing, with a glass transition occurring at relatively low temperatures; thus, the molecular effect provided by the glass transition directly affects the precision and accuracy of 3D printing [116]. It is worth noting that inks based on starch gels can present an incomplete gel network at low printing temperatures, as the free macromolecular chains are limited, and the vast majority remain arranged in the amorphous and crystalline portions of the granules. However, at high printing temperatures, the free macromolecules exhibit greater long-range disorder due to thermal energy, breaking hydrogen bonds and other intermolecular forces, and reducing the mechanical strength and structure of the gel [17]. Therefore, an ink with good thermal stability is one that maintains print quality in proportion to the magnitude of the temperature used.

7.6 Starch Used as Bio-ink for 3D Printed Food Application

7.6.1 Tailored Food

The 3D printing technology has gained attention as an excellent technique for food production, as it allows the development of materials with customized sizes and shapes, as well as different textures, flavors, and nutritional profiles according to the additives used [49]. In this context, starch has been explored as a raw material for use as an ink in 3D printing applications. Starch is a prominent carbohydrate in human diets and is found in the nature in various botanical sources. It is also a relatively inexpensive source. Moreover, starch is already employed as a thickening and gelling agent in food formulations [52], which renders its use for the production of 3D printed foods more favorable and more acceptable to the consumers. Furthermore, the production of food through 3D printing technology has the incredible potential and advantage of enabling the production of personalized foods for people who have difficulties in chewing and swallowing, a condition known as dysphagia [19, 117]. Dysphagia, a condition that affects a significant proportion of the elderly population, is increasingly recognized as a pressing healthcare concern, potentially resulting in malnutrition, dehydration, and decreased muscle mass [118].

In this regard, Guedes et al. [19] produced 3D printed modified wheat starch-based materials with a potential for creating food products tailored to people with dysphagia. The gels were generated with improved printability, weaker consistency, and foods characterized at levels 5 and 6 by the International Dysphagia Diet Standardization Initiative (IDDSI). Wang et al. [20] observed the effect of different types of starch (wheat, corn, rice, potato, sweet potato, cassava, mung bean and pea) to structure composite emulsion gels with fibrous whey protein (CEGs). In this work, rice and cassava starches were better suited for structuring soft-textured materials, with a IDDSI categorized as level 5. Huang et al. [84] obtained improvements in the printability and self-support of 3D printed materials from dry heated modified starches (white quinoa and brown rice), also improving resistance to enzymatic degradation and the digestibility of the treated samples.

Furthermore, starch can also be used together with nutrients or other components that can add nutritional value to the materials produced via 3D printing, highlighting another advantage of this technology: the ability to produce foods with personalized nutritional value [49]. Jo et al. [119] explored the effect of three types of starch (waxy corn, tapioca and potato) in 3D printed functional snacks with addition of 20% w/w of red ginseng. The authors noted that the highest shear modulus and the best dimensional stability and printing resolution were obtained with the use of high amylose potato starch (at 10% w/w). At the end, these snacks showed a relative hardness between pastries and dark chocolate (20.1 N and 546 N, respectively). Montoya et al. [120] evaluated the potential of starch-mango and starch-arabinoxylan blends as raw materials for 3D printed food. The results showed that high starch content in the blends makes them more suitable for 3D food printing, producing materials with solid-like behavior, improved structural integrity, and high thixotropy, making them a desirable material for the 3D printing process.

7.6.2 Functional Food

Functional foods are those that have specific activities capable of modulating bodily functions, in addition to their basic nutritional value [121]. They often contain active substances that can help in the prevention and management of chronic diseases, support metabolic functions, enhance overall well-being, or meet the needs of specific populations [122, 123]. The advent of 3D printing in the food industry has enabled the creation of highly customized functional food products to enhance their actives, rheological and physicochemical properties.

By utilizing starch-based bio-inks, it is possible to design and manufacture foods that are not only tailored to individual dietary preferences but also enhanced with functional ingredients. The versatile properties of starch (i.e., ability to gelation, capacity to formed 3D open porous structure after gelation, non-toxicity, and biocompatibility with many other substances) [124], allows it can act as an excellent carrier for active substances, improving distribution and effective bioaccessibility after consumption. Furthermore, starch-based bio-inks can produce functional foods with texture and swallowability properties desirable for specific groups [125].

Several innovative examples demonstrate the potential of starch-based bio-inks to produce 3D printed functional foods, includes foods enriched with vitamins [126], minerals [123], probiotics [125], fibers [127], proteins [128, 129], antioxidants [130, 131], and other active substances. In these cases, the use of starch was interesting to increase the apparent viscosity of the bio-ink and improve the texture and mechanical properties of the formulations, in addition to carrying some of these active substances. For example, researchers explored the addition of strawberry tree fruits extract, rich in polyphenols, to wheat and corn starch-based bio-inks for 3D printing of functional foods with antioxidant and antimicrobial activities [130]. The strawberry tree fruits extract protected the organism from the harmful effects of intracellular reactive oxygen species on DNA and demonstrated to be an efficient antibacterial against *Escherichia coli* and *Staphylococcus aureus*. The physicochemical and biological properties of the 3D printed formulations were dependent on the type and content of starch and proved to be valuable to produce functional foods with high biological potential.

Personalized functional foods for the elderly, children and dysphagic patients can also be developed seeking digestive and/or textured food solutions for these groups. Starch-rich brown rice bio-inks were used to modulate and encapsulate probiotics (*Lactobacillus bulgaricus*, *Lactobacillus paracasei*, *Bifidobacterium longum*, *Streptococcus thermophilus*, and *Lactiplantibacillus plantarum*) by 3D printing these formulations for the elderly people [125]. Some bio-inks have been categorized as having easy swallowing characteristics. There was modulation of nutrients and flavor, producing fatty acids and improving starch digestion. Gels based on native and modified cassava starch, kappa-carrageenan and milk were also developed with a focus on elderly and hep dysphagia peoples. The 3D printed gels presented the desired texture and high bioaccessibility of the minerals present in milk (calcium, magnesium, potassium, and phosphorus) [123].

7.6.3 Packaging Materials

In the field of 3D printing food, the versatility of starch makes it suitable not only as a base material for the production of inks for 3D-printed food but also as a matrix for the production of biodegradable films to be used as packaging materials for food products [49]. In this context, increased attention has been given to the development of active and smart packaging films with antibacterial, antioxidant, and barrier properties [132].

In this sense, Leaw et al. [133] produced antibacterial food packaging based on 3D-printed corn starch-gelatin films, with different concentrations of glycerol (1–5% w/v) and hawthorn berry extract (2, 4, 6, 8, and 10% w/v). In the optimized conditions (4% glycerol and 6% of hawthorn berry extract), the authors obtained films with improved thickness, water solubility, and elongation at break when compared to the control. It also inhibited the growth of Pseudomonas aeruginosa. Dermol et al. [134] produced a homogeneous 3D-printed foils of modified cationic starch and bacterial nanocellulose (optimized condition with a proportion of 50/50), with higher water contact angles and tensile strength and elastic modulus comparable to various commercially used synthetic foils [132].

In addition, Bao et al. [104] developed a highly responsive 3D printed colorimetric film made of potato starch as matrix incorporated with blueberry anthocyanin-phycocyanin (BA-PC), as an indicator, and chondroitin sulfate as a co-pigment. The films presented a smooth and homogeneous gel structure, with high sensitivity to ammonia (at concentration of 20% v/v) and high antioxidant activity, responding efficiently to the freshness of salmon and beef in the performed tests. Li et al. [135] used 3D printing for production of sandwich-type indicator films constituted of cassava starch as a protective layer, lemongrass essential oil to add antibacterial and antioxidant properties, mulberry anthocyanin as colorimetric sensor, and chitosan. According to the authors, the films showed good antioxidant and bacteriostatic properties, and excellent colorimetric characteristics.

Daza et al. [136] evaluated the potential use of a nonconventional starch from chachafruto as source for 3D-printed films. The authors obtained materials with good thermal stability, a reduced crystallinity, good elasticity and high consistency, high transparency, and good barrier properties.

7.7 Perspectives in the Use of Starch in 3D Food Printing

Emerging technologies and interdisciplinary approaches continue to expand the horizons of starch utilization in 3D food printing, driving innovation in personalized nutrition, sustainable food systems, and food design.

Starch's ability to form gels, control viscosity, and stabilize emulsions makes it a cornerstone for creating printable food inks. Modified starches, such as pregelatinized or cross-linked variants, offer enhanced rheological properties, enabling consistent flow behavior and shape retention during printing. These modifications also expand the scope of textures and functionalities achievable in printed foods [17]. Furthermore, starch's natural compatibility with other biomolecules, such as proteins and lipids, allows for the development of composite formulations tailored to specific nutritional and sensory profiles.

3D food printing using starch-based materials opens avenues for creating customized foods that cater to individual dietary needs, preferences, and health goals. Starch can act as a carrier for bioactive compounds, vitamins, and minerals, enabling the development of fortified and functional foods [27]. For example, the integration of resistant starch could enhance the nutritional profile by promoting gut health and reducing the glycemic index of printed products. The precision of 3D printing also allows for the creation of intricate structures that optimize flavor release and texture perception, enhancing the consumer experience.

The use of starch in 3D food printing aligns with sustainability goals by promoting the utilization of plant-based and renewable resources. Starch derived from underutilized crops, such as cassava, potato, or green bananas, offers an opportunity to diversify raw material sources while reducing agricultural waste [23]. Incorporating starch-based inks into circular economy models could also facilitate the development of biodegradable food packaging or edible structures, addressing environmental concerns related to single-use plastics.

Despite its advantages, several challenges remain in the use of starch for 3D food printing. The material's sensitivity to temperature, moisture, and shear stress can affect the stability and reproducibility of printed structures. Ongoing research focuses on improving the thermal stability and mechanical strength of starch-based inks through chemical and enzymatic modifications [137]. Additionally, integrating advanced printing technologies, such as multi-material extrusion and hybrid printing systems, could further enhance the performance and versatility of starch in food printing.

The growing interest in plant-based diets, coupled with the demand for personalized and convenient foods, positions starch as a key ingredient in the evolving 3D food printing market. Consumers are increasingly seeking novel and sustainable food products, and starch-based printing technologies are well-poised to address these preferences. Collaboration between researchers, food technologists, and the food industry will be essential to accelerate the adoption of starch in commercial 3D food printing applications.

In conclusion, starch presents immense potential as a sustainable and functional material for 3D food printing. Advances in material science, processing technologies, and consumer acceptance will shape its future, paving the way for a new era of personalized, sustainable, and innovative food solutions.

7.8 Conclusions

Starch emerges as a vital material for 3D food printing, offering unparalleled versatility and functionality. Its natural properties, such as gelation and viscosity control, support the precise extrusion and structural integrity needed for layer-by-layer food creation. Innovations in modifying starch, whether physical, chemical, or enzymatic, have expanded its potential, enabling applications ranging from functional foods to sustainable packaging.

Moreover, the integration of starch in 3D printing aligns seamlessly with global demands for sustainable and personalized food solutions. As research continues to refine its printability and explore new applications, starch holds the promise of transforming the food industry into a more adaptive, health-focused, and environmentally conscious sector. By leveraging advancements in additive manufacturing and interdisciplinary collaboration, starch-based technologies are poised to unlock a new era in food design and production.

Acknowledgments We would like to thank the São Paulo Research Foundation (FAPESP, Brazil) for funding the projects 2024/10994-8, 2020/08727-0, 2023/17512-6, and 2023/03995-5. The National Council for Scientific and Technological Development (CNPq, Brazil, grant #162334/2021-4). The "Coordenação de Aperfeiçoamento de Pessoal de Nível Superior - Brasil (CAPES)" (88887.636998/2021-00). The CAPES/Brazil and COFECUB/France for funding the project Ph 1006/23 (CAPES code 001).

Conflicts of Interest The authors declare no conflict of interest.

References

1. Taneja A, Sharma R, Ayush K, Sharma A, Mousavi Khaneghah A, Regenstein JM, et al. Innovations and applications of 3-D printing in food sector. Int J Food Sci Technol. 2022;57(6):3326–32.
2. Jiao X, Ren G, Law CL, Li L, Cao W, Luo Z, et al. Novel strategy for optimizing of corn starch-based ink food 3D printing process: printability prediction based on BP-ANN model. Int J Biol Macromol. 2024;276:133921.
3. Bassan de Moraes MJ, Nagata EY, Felício Peres Duran AJ, Rossignolo JA. Alkali activated materials applied in 3D printing construction: a review. Heliyon. 2024;10(5):e26696.
4. Boretti A. A techno-economic perspective on 3D printing for aerospace propulsion. J Manuf Process. 2024;109:607–14.
5. Yang L, Meng J, Xue T, Wang Y, Shi G, Gao X, et al. Application of 3D printing cellulose fabrics based on cotton fibers in the textile and fashion industry. Addit Manuf. 2024;81:104000.
6. Kulkarni VR, Saha T, Raj Giri B, Lu A, Das SC, Maniruzzaman M. recent advancements in pharmaceutical 3D printing industry. J Drug Deliv Sci Technol. 2024;100:106072.
7. Mohseni Y, Mohseni M, Suresh S, Riotto M, Jaggessar A, Little JP, et al. Investigating impacts of FDM printing parameters and geometrical features on void formation in 3D printed automotive components. Mater Today Proc. 2023; https://doi.org/10.1016/j.matpr.2023.06.078.
8. Tracey CT, Predeina AL, Krivoshapkina EF, Kumacheva E. A 3D printing approach to intelligent food packaging. Trends Food Sci Technol. 2022;127:87–98.
9. Chen J, Zhou H, Fan Y, Gao G, Ying Y, Li J. 3D printing for bone repair: coupling infection therapy and defect regeneration. Chem Eng J. 2023;471:144537.
10. Sartori de AGO, Saliba ASMC, Bitencourt BS, Guedes JS, Torres LCR, Alencar de SM, et al. Anthocyanin bioaccessibility and anti-inflammatory activity of a grape-based 3D printed food for dysphagia. Innovative Food Sci Emerg Technol 2023;84:103289.
11. Zhong Y, Wang B, Lv W, Li G, Lv Y, Cheng Y. Egg yolk powder-starch gel as novel ink for food 3D printing: rheological properties, microstructure and application. Innovative Food Sci Emerg Technol. 2024;91:103545.

12. Cheng Y, Yuqing H, Xiao L, Gao W, Kang X, Sui J, et al. Impact of starch amylose and amylopectin on the rheological and 3D printing properties of corn starch. Int J Biol Macromol. 2024;278:134403.
13. Malki MKS, Wijesinghe JAAC, Ratnayake RHMK, Thilakarathna GC. Characterization of arrowroot (Maranta arundinacea) starch as a potential starch source for the food industry. Heliyon. 2023;9(9):e20033.
14. Maniglia BC, Lima DC, Matta Junior MD, Le-Bail P, Le-Bail A, Augusto PED. Hydrogels based on ozonated cassava starch: effect of ozone processing and gelatinization conditions on enhancing 3D-printing applications. Int J Biol Macromol. 2019;138:1087–97.
15. Maniglia BC, Castanha N, Le-Bail P, Le-Bail A, Augusto PED. Starch modification through environmentally friendly alternatives: a review. Crit Rev Food Sci Nutr. 2021;61(15):2482–505.
16. Lima DC, Villar J, Castanha N, Maniglia BC, Matta Junior MD, Duarte Augusto PE. Ozone modification of arracacha starch: effect on structure and functional properties. Food Hydrocoll. 2020;108:106066.
17. Rong L, Chen X, Shen M, Yang J, Qi X, Li Y, et al. The application of 3D printing technology on starch-based product: a review. Trends Food Sci Technol. 2023;134:149–61.
18. do Val Siqueira L, Arias CILF, Maniglia BC, Tadini CC. Starch-based biodegradable plastics: methods of production, challenges and future perspectives. Curr Opin Food Sci. 2021;38:122–30.
19. Guedes JS, Bitencourt BS, Augusto PED. Modification of maize starch by dry heating treatment (DHT) and its use as gelling ingredients in fruit-based 3D-printed food for dysphagic people. Food Biosci. 2023;56:103310.
20. Wang Z, Chen F, Deng Y, Tang X, Li P, Zhao Z, et al. Texture characterization of 3D printed fibrous whey protein-starch composite emulsion gels as dysphagia food: a comparative study on starch type. Food Chem. 2024;458:140302.
21. Wang Y, Zhao R, Liu W, Zhao R, Liu Q, Hu H. Effect of twin-screw extrusion pretreatment on starch structure, rheological properties and 3D printing accuracy of whole potato flour and its application in dysphagia diets. Int J Biol Macromol. 2024;278:134796.
22. Wu Y, Li Z, Yang L, Zheng H, Xue C. Construction of 3D printed salmon fillet simulants: improving printing performance by blending corn starch and flaxseed oil with pea protein and post-printing texturization via transglutaminase. Food Hydrocoll. 2024;155:110242.
23. Gulzar S, Narciso JO, Elez-Martínez P, Martín-Belloso O, Soliva-Fortuny R. Recent developments in the application of novel technologies for the modification of starch in light of 3D food printing. Curr Opin Food Sci. 2023;52:101067.
24. Kafle A, Luis E, Silwal R, Pan HM, Shrestha PL, Bastola AK. 3D/4D printing of polymers: Fused Deposition Modelling (FDM), Selective Laser Sintering (SLS), and Stereolithography (SLA). Polymers (Basel). 2021;13(18):3101.
25. Chen Y, McClements DJ, Peng X, Chen L, Xu Z, Meng M, et al. Starch as edible ink in 3D printing for food applications: a review. Crit Rev Food Sci Nutr. 2024;64(2):456–71.
26. Sun J, Zhou W, Huang D, Fuh JYH, Hong GS. An overview of 3D printing technologies for food fabrication. Food Bioprocess Technol. 2015;8(8):1605–15.
27. Sun J, Zhou W, Yan L, Huang D, Lin L, ya. Extrusion-based food printing for digitalized food design and nutrition control. J Food Eng. 2018;220:1–11.
28. Mantihal S, Kobun R, Lee BB. 3D food printing of as the new way of preparing food: a review. Int J Gastron Food Sci. 2020;22:100260.
29. Guo CF, Zhang M, Bhandari B. A comparative study between syringe-based and screw-based 3D food printers by computational simulation. Comput Electron Agric. 2019;162:397–404.
30. Maniglia BC, Lima DC, Matta Junior MD, Le-Bail P, Le-Bail A, Augusto PED. Preparation of cassava starch hydrogels for application in 3D printing using dry heating treatment (DHT): a prospective study on the effects of DHT and gelatinization conditions. Food Res Int [Internet]. 2020;128:108803. Available from: https://linkinghub.elsevier.com/retrieve/pii/S0963996919306891.

31. Bugarin-Castillo Y, Rando P, Clabaux M, Moulin G, Ramaioli M. 3D printing to modulate the texture of starch-based food. J Food Eng. 2023;350:111499.
32. Xu M, Xu Y, Ji S, Amrouche AT, Li Y, Zhou Z, et al. Investigation of the 3D printability of modified starch-based inks and their formation mechanism: application in ice cream. Food Hydrocoll. 2024;154:110038.
33. Li Z, Wang Z, Gan X, Fu D, Fei G, Xia H. Selective laser sintering 3D printing: a way to construct 3D electrically conductive segregated network in polymer matrix. Macromol Mater Eng. 2017;302(11):1700211.
34. Awad A, Fina F, Goyanes A, Gaisford S, Basit AW. 3D printing: principles and pharmaceutical applications of selective laser sintering. Int J Pharm. 2020;586:119594.
35. Jonkers N, van Dijk WJ, Vonk NH, van Dommelen JAW, Geers MGD. Anisotropic mechanical properties of selective laser sintered starch-based food. J Food Eng. 2022;318:110890.
36. Holland S, Foster T, MacNaughtan W, Tuck C. Design and characterisation of food grade powders and inks for microstructure control using 3D printing. J Food Eng. 2018;220:12–9.
37. Ziaee M, Crane NB. Binder jetting: a review of process, materials, and methods. Addit Manuf. 2019;28:781–801.
38. Miyanaji H, Rahman KM, Da M, Williams CB. Effect of fine powder particles on quality of binder jetting parts. Addit Manuf. 2020;36:101587.
39. Mostafaei A, Elliott AM, Barnes JE, Li F, Tan W, Cramer CL, et al. Binder jet 3D printing—Process parameters, materials, properties, modeling, and challenges. Prog Mater Sci. 2021;119:100707.
40. Jiang Q, Zhang M, Mujumdar AS. Novel evaluation technology for the demand characteristics of 3D food printing materials: a review. Crit Rev Food Sci Nutr. 2022;62(17):4669–83.
41. Zhu S, Vazquez Ramos P, Heckert OR, Stieger M, van der Goot AJ, Schutyser M. Creating protein-rich snack foods using binder jet 3D printing. J Food Eng. 2022;332:111124.
42. Tetlow IJ, Bertoft E. A review of starch biosynthesis in relation to the building block-backbone model. Int J Mol Sci. 2020;21(19):7011.
43. Chen S, Qin L, Chen T, Yu Q, Chen Y, Xiao W, et al. Modification of starch by polysaccharides in pasting, rheology, texture and in vitro digestion: a review. Int J Biol Macromol. 2022;207:81–9.
44. Seung D. Amylose in starch: towards an understanding of biosynthesis, structure and function. New Phytol. 2020;228(5):1490–504.
45. Wang Y, Ou X, Al-Maqtari QA, He HJ, Othman N. Evaluation of amylose content: Structural and functional properties, analytical techniques, and future prospects. Food Chem X. 2024;24:101830.
46. Alcázar-Alay SC, Meireles MAA. Physicochemical properties, modifications and applications of starches from different botanical sources. Food Sci Technol (Campinas) [Internet]. 2015;35(2):215–36. Available from: http://www.scielo.br/scielo.php?script=sci_arttext&pid=S0101-20612015000200215&lng=en&nrm=iso&tlng=en.
47. Magallanes-Cruz PA, Flores-Silva PC, Bello-Perez LA. Starch structure influences its digestibility: a review. J Food Sci. 2017;82(9):2016–23.
48. Cheng Y, Liang K, Chen Y, Gao W, Kang X, Li T, et al. Effect of molecular structure changes during starch gelatinization on its rheological and 3D printing properties. Food Hydrocoll. 2023;137:108364.
49. Chen Y, McClements DJ, Peng X, Chen L, Xu Z, Meng M, et al. Starch as edible ink in 3D printing for food applications: a review. In: Critical reviews in food science and nutrition, vol. 64. Taylor and Francis Ltd.; 2024. p. 456–71.
50. Mohamed AA, Alqah H, Alamri MS, Hussain S, Qasem AA, Ibraheem MI, et al. Physicochemical properties of enzymatically modified starches. Processes. 2021;9(12), 2251.
51. Ji S, Xu T, Li Y, Li H, Zhong Y, Lu B. Effect of starch molecular structure on precision and texture properties of 3D printed products. Food Hydrocoll. 2022;125:107387.

52. Maniglia BC, Lima DC, da Matta JM, Oge A, Le-Bail P, Augusto PED, et al. Dry heating treatment: a potential tool to improve the wheat starch properties for 3D food printing application. Food Res Int. 2020;137:109731.
53. Maleki S, Aarabi A, Far FA, Dizaji HZ. Heat moisture treatment and ultrasound-induced hydrothermal wheat starch modification: techno-functional, microstructural and quality 3D printed characteristics. Int J Biol Macromol. 2024;276:133992.
54. Maniglia BC, Pataro G, Ferrari G, Augusto PED, Le-Bail P, Le-Bail A. Pulsed electric fields (PEF) treatment to enhance starch 3D printing application: effect on structure, properties, and functionality of wheat and cassava starches. Innov Food Sci Emerg Technol. 2021;68:102602.
55. Larrea-Wachtendorff D, Tabilo-Munizaga G, Ferrari G. Potato starch hydrogels produced by high hydrostatic pressure (HHP): a first approach. Polymers (Basel). 2019;11(10):1673.
56. Sharma V, Kaur M, Singh K, Kaur S. Barnyard millet starch cross-linked at varying levels by sodium trimetaphosphate (STMP): film forming , physico-chemical, pasting and thermal properties. Carbohydr Polym Technol Appl [Internet]. 2021;2:100161. Available from: https://doi.org/10.1016/j.carpta.2021.100161.
57. Wei W, Wu M, Shang N, Mu R, Zhang T, Xu H. Low temperature preparation of high-performance citric acid crosslinked starch gels with adjustable properties based on cationic esterification synergistic crosslinking. Food Hydrocoll. 2023;145:109098.
58. Park HR, Kang J, Rho SJ, Kim YR. Structural and physicochemical properties of enzymatically modified rice starch as influenced by the degree of enzyme treatment. J Carbohydr Chem. 2020;39:250–66.
59. Li X, Wang Y, Lee BH, Li D. Reducing digestibility and viscoelasticity of oat starch after hydrolysis by pullulanase from Bacillus acidopullulyticus. Food Hydrocoll. 2018;75:88–94.
60. Jayarathna S, Hofvander P, Péter-Szabó Z, Andersson M, Andersson R. GBSS mutations in an SBE mutated background restore the potato starch granule morphology and produce ordered granules despite differences to native molecular structure. Carbohydr Polym. 2024;331:121860.
61. Sinhmar A, Pathera AK, Sharma S, Nehra M, Thory R, Nain V. Impact of various modification methods on physicochemical and functional properties of starch: a review. Starch/Staerke. 2023;75(1–2):1–14.
62. Zheng B, Qiu Z, Liu Z, Chen L. Pre-dry heat treatment alters the structure and ultimate in vitro digestibility of wheat starch-lipids complex in hot-extrusion 3D printing. Carbohydr Polym. 2024;334:122026.
63. Maniglia BC, Lima DC, Matta Junior MD, Le-Bail P, Le-Bail A, Augusto PED. Preparation of cassava starch hydrogels for application in 3D printing using dry heating treatment (DHT): a prospective study on the effects of DHT and gelatinization conditions. Food Res Int [Internet]. 2020;128:108803. Available from: https://doi.org/10.1016/j.foodres.2019.108803.
64. Sponchiado PAI, de Melo MT, Bitencourt BS, Guedes JS, Tapia-Blácido DR, Augusto PED, et al. Clean modification of potato starch to improve 3D printing of potential bone bioscaffolds. Emerg Mater. 2024:1–14.https://doi.org/10.1007/s42247-024-00673-6
65. Li ZY, Hua CL, Yong YX, Hui JZ, Hao QW, Deng LP, et al. Effects of repeated and continuous dry heat treatments on the physicochemical and structural properties of quinoa starch. Food Hydrocoll. 2021;113:106532.
66. Liu S, Liu H, Gao S, Guo S, Zhang C. Dry heating affects the multi-structures, physicochemical properties, and in vitro digestibility of blue highland barley starch. Front Nutr. 2023;10:1191391.
67. Wang Q, Li L, Zheng X. Recent advances in heat-moisture modified cereal starch: structure, functionality and its applications in starchy food systems. Food Chem [Internet]. 2021;344:128700. Available from: http://www.ncbi.nlm.nih.gov/pubmed/33248839.
68. Zheng L, Zhang Q, Yu X, Luo X, Jiang H. Effect of annealing and heat-moisture pretreatment on the quality of 3D-printed wheat starch gels. Innov Food Sci Emerg Technol. 2023;84:103274.

69. Castanha N, Santos DN, Cunha RL, Augusto PED. Properties and possible applications of ozone-modified potato starch. Food Res Int. 2019;116:1192–201.
70. Haq F, Yu H, Wang L, Teng L, Haroon M, Khan RU, et al. Advances in chemical modifications of starches and their applications. Carbohydr Res [Internet]. 2019;476(February):12–35. Available from: https://doi.org/10.1016/j.carres.2019.02.007.
71. Maniglia BC, Castanha N, Le-Bail P, Le-Bail A, Augusto PED. Starch modification through environmentally friendly alternatives: a review. Crit Rev Food Sci Nutr [Internet]. 2021;61(15):2482–505. Available from: https://doi.org/10.1080/10408398.2020.1778633.
72. Lima DC, Castanha N, Maniglia BC, Matta Junior MD, La Fuente CIA, Augusto PED. Ozone processing of cassava starch. Ozone Sci Eng. 2021;43(1):60–77.
73. Gerezgiher AG, Szabó T. Crosslinking of starch using citric acid. In: Journal of Physics: Conference Series. Institute of Physics; 2022.
74. Nakkala K, Godiyal S, Ettaboina SK, Laddha KS. Chemical modifications of turmeric starch by oxidation, phosphorylation, and succinylation. Starch/Staerke. 2022;2200053:1–9.
75. Punia Bangar S, Ashogbon AO, Singh A, Chaudhary V, Whiteside WS. Enzymatic modification of starch: a green approach for starch applications. In: Carbohydrate polymers, vol. 287. Elsevier Ltd; 2022.
76. Shah A, Masoodi FA, Gani A, Ashwar B. Dual enzyme modified oat starch: structural characterisation, rheological properties, and digestibility in simulated GI tract. Int J Biol Macromol. 2018;106:140–7.
77. Geng DH, Tang N, Gan J, Cheng Y. Two-step modification of pullulanase and transglucosidase: a novel way to improve the gel strength and reduce the digestibility of rice starch. Int J Biol Macromol. 2024;266:130992.
78. Rong S, Chen S, Su P, Tang H, Jia M, Xia Y, et al. Postsynthetic modification of metal–organic frameworks by vapor-phase grafting. Inorg Chem. 2021;60(16):11745–9.
79. Farooq S, Ahmad MI, Zhang Y, Chen M, Zhang H. Preparation, characterization and digestive mechanism of plant-derived oil bodies based oleogels structured by chitosan and vanillin. Food Hydrocoll. 2023;136:108247.
80. Liu S, Xiao Y, Shen M, Zhang X, Wang W, Xie J. Effect of sodium carbonate on the gelation, rheology, texture and structural properties of maize starch-Mesona chinensis polysaccharide gel. Food Hydrocoll. 2019;87:943–51.
81. Cai Q, Zhong Y, Xu M, Huang Q, Lu X. 3D printed high oil custard cream: effects of whey protein isolate, hydroxypropylated starch and carrageenan on physicochemical properties and printing performance. LWT. 2022;156:113039.
82. Cui Y, Li C, Guo Y, Liu X, Zhu F, Liu Z, et al. Rheological & 3D printing properties of potato starch composite gels. J Food Eng. 2022;313:110756.
83. Chen Y, Zhang M, Sun Y, Phuhongsung P. Improving 3D/4D printing characteristics of natural food gels by novel additives: a review. Food Hydrocoll. 2022;123:107160.
84. Huang J, Zhang M, Mujumdar AS, Li C. Modulation of starch structure, swallowability and digestibility of 3D-printed diabetic-friendly food for the elderly by dry heating. Int J Biol Macromol. 2024;264:130629.
85. Morell P, Fiszman SM, Varela P, Hernando I. Hydrocolloids for enhancing satiety: relating oral digestion to rheology, structure and sensory perception. Food Hydrocoll. 2014;41:343–53.
86. Wang K, Li M, Han Q, Fu R, Ni Y. Inhibition of α-amylase activity by insoluble and soluble dietary fibers from kiwifruit (Actinidia deliciosa). Food Biosci. 2021;42:101057.
87. Zhang B, Qiao D, Zhao S, Lin Q, Wang J, Xie F. Starch-based food matrices containing protein: recent understanding of morphology, structure, and properties. Trends Food Sci Technol. 2021;114:212–31.
88. Guo C, Zhang M, Devahastin S. Improvement of 3D printability of buckwheat starch-pectin system via synergistic Ca2+-microwave pretreatment. Food Hydrocoll. 2021;113:106483.
89. Dankar I, Pujolà M, El Omar F, Sepulcre F, Haddarah A. Impact of mechanical and microstructural properties of potato puree-food additive complexes on extrusion-based 3D printing. Food Bioprocess Technol. 2018;11(11):2021–31.

90. Lille M, Nurmela A, Nordlund E, Metsä-Kortelainen S, Sozer N. Applicability of protein and fiber-rich food materials in extrusion-based 3D printing. J Food Eng. 2018;220:20–7.
91. Liu Y, Yi S, Sameen DE, Hossen MA, Dai J, Li S, et al. Designing and utilizing 3D printed chitosan/halloysite nanotubes/tea polyphenol composites to maintain the quality of fresh blueberries. Innovative Food Sci Emerg Technol. 2021;74:102808.
92. Yang C, Zhong F, Douglas Goff H, Li Y. Study on starch-protein interactions and their effects on physicochemical and digestible properties of the blends. Food Chem. 2019;280:51–8.
93. Mu X, Agostinacchio F, Xiang N, Pei Y, Khan Y, Guo C, et al. Recent advances in 3D printing with protein-based inks. Prog Polym Sci. 2021;115:101375.
94. Yang F, Zhang M, Liu Y. Effect of post-treatment microwave vacuum drying on the quality of 3D-printed mango juice gel. Dry Technol. 2019;37:1757–65.
95. Kim Y, Kim HW, Park HJ. Effect of pea protein isolate incorporation on 3D printing performance and tailing effect of banana paste. LWT. 2021;150:111916.
96. Riantiningtyas RR, Sager VF, Chow CY, Thybo CD, Bredie WLP, Ahrné L. 3D printing of a high protein yoghurt-based gel: effect of protein enrichment and gelatine on physical and sensory properties. Food Res Int. 2021;147:110517.
97. Hussain S, Arora VK, Malakar S. Formulation of protein-enriched 3D printable food matrix and evaluation of textural, rheological characteristics, and printing stability. J Food Process Preserv. 2021;45(2):e15182.
98. Chuanxing F, Qi W, Hui L, Quancheng Z, Wang M. Effects of pea protein on the properties of potato starch-based 3D printing materials. Int J Food Eng. 2018;14(3):20170297.
99. Ning X, Devahastin S, Wang X, Wu N, Liu Z, Gong Y, et al. Understanding 3D food printing through computer simulation and extrusion force analysis. J Food Eng. 2024;370:111972.
100. Outrequin TCR, Gamonpilas C, Sreearunothai P, Deepaisarn S, Siriwatwechakul W. Machine learning assisted evaluation of the filament spreading during extrusion-based 3D food printing: impact of the rheological and printing parameters. J Food Eng. 2024;381:112166.
101. Nijdam JJ, LeCorre-Bordes D, Delvart A, Schon BS. A rheological test to assess the ability of food inks to form dimensionally stable 3D food structures. J Food Eng. 2021;291:110235.
102. Park BR, No J, Oh H, Park CS, You KM, Chewaka LS. Exploring puffed rice as a novel ink for 3D food printing: rheological characterization and printability analysis. J Food Eng. 2025;387:112313.
103. Wang L, Zhang M, Bhandari B, Yang C. Investigation on fish surimi gel as promising food material for 3D printing. J Food Eng. 2018;220:101–8.
104. Bao Y, Wang M, Li J, Yang X, Wen B, Wang L, et al. A starch-based 3D printed intelligent colorimetric film co-loaded natural pigments for visualizing food freshness: effect of nozzle size on gel structure formation. Food Hydrocoll. 2024;155:110218.
105. Ma S, Bao Y, Xu M, Yu X, Jiang H. Effect of 3D printing and traditional molding on phenolic compounds and antioxidant activity in steamed bread. Food Chem. 2024;454:139699.
106. Krishnaraj P, Anukiruthika T, Choudhary P, Moses JA, Anandharamakrishnan C. 3D extrusion printing and post-processing of fibre-rich snack from indigenous composite flour. Food Bioprocess Technol. 2019;12(10):1776–86.
107. Ahmadzadeh S, Lenie MDR, Mirmahdi RS, Ubeyitogullari A. Designing future foods: harnessing 3D food printing technology to encapsulate bioactive compounds. Crit Rev Food Sci Nutr. 2023;26:1–17.
108. Ahmadzadeh S, Ubeyitogullari A. Lutein encapsulation into dual-layered starch/zein gels using 3D food printing: Improved storage stability and in vitro bioaccessibility. Int J Biol Macromol. 2024;266:131305.
109. Derossi A, Caporizzi R, Azzollini D, Severini C. Application of 3D printing for customized food. A case on the development of a fruit-based snack for children. J Food Eng. 2018;220:65–75.
110. Derossi A, Caporizzi R, Azzollini D, Severini C. Application of 3D printing for customized food. A case on the development of a fruit-based snack for children. J Food Eng

[Internet]. 2018;220:65–75. Available from: https://linkinghub.elsevier.com/retrieve/pii/S0260877417302133.
111. Huan CX, Zhang M, Xiu TX, Mujumdar AS. Internal structure design for improved shape fidelity and crispness of 3D printed pumpkin-based snacks after freeze-drying. Food Res Int. 2022;157:111220.
112. Liu Q, Zhang N, Wei W, Hu X, Tan Y, Yu Y, et al. Assessing the dynamic extrusion-based 3D printing process for power-law fluid using numerical simulation. J Food Eng [Internet]. 2020;275:109861. Available from: https://linkinghub.elsevier.com/retrieve/pii/S0260877419305047.
113. Ma Y, Schutyser MAI, Boom RM, Zhang L. Thermographic and rheological characterization of viscoelastic materials for hot-extrusion 3D food printing. Innovative Food Sci Emerg Technol. 2022;81:103135.
114. Chieregato Maniglia B, Carregari Polachini T, Norwood EA, Le-Bail P, Le-Bail A. Thermal technologies to enhance starch performance and starchy products. Curr Opin Food Sci. 2021;40:72–80.
115. Bhandari BR, Howes T. Implication of glass transition for the drying and stability of dried foods. J Food Eng. 1999;40(1–2):71–9.
116. Liu Z, Zhang M, Bhandari B, Wang Y. 3D printing: printing precision and application in food sector. Trends Food Sci Technol. 2017;69:83–94.
117. de Sartori AGO, Saliba ASMC, Bitencourt BS, Guedes JS, Torres LCR, de Alencar SM, et al. Anthocyanin bioaccessibility and anti-inflammatory activity of a grape-based 3D printed food for dysphagia. Innov Food Sci Emerg Technol. 2023;84:103289.
118. Chao C, Nam HK, Park HJ, Kim HW. Potentials of 3D printing in nutritional and textural customization of personalized food for elderly with dysphagia. In: Applied biological chemistry, vol. 67. Springer Science and Business Media B.V.; 2024.
119. Jo GH, Lim WS, Kim HW, Park HJ. Post-processing and printability evaluation of red ginseng snacks for three-dimensional (3D) printing 3D printing of snacks according to starch type. Food Biosci. 2021;42:101094.
120. Montoya J, Medina J, Molina A, Gutiérrez J, Rodríguez B, Marín R. Impact of viscoelastic and structural properties from starch-mango and starch-arabinoxylans hydrocolloids in 3D food printing. Addit Manuf. 2021;39:101891.
121. Wu R, Jiang J, An F, Ma X, Wu J. Research progress of 3D printing technology in functional food, powering the future of food. Trends Food Sci Technol [Internet]. 2024 Jul 1 [cited 2024 Sep 15];149:104545. Available from: https://linkinghub.elsevier.com/retrieve/pii/S0924224424002218.
122. Portanguen S, Tournayre P, Sicard J, Astruc T, Mirade PS. Toward the design of functional foods and biobased products by 3D printing: a review. Trends Food Sci Technol [Internet]. 2019 Apr 1 [cited 2024 Sep 15];86:188–98. Available from: https://linkinghub.elsevier.com/retrieve/pii/S092422441830308X.
123. Bitencourt BS, Guedes JS, Saliba ASMC, Sartori AGO, Torres LCR, Amaral JEPG, et al. Mineral bioaccessibility in 3D printed gels based on milk/starch/κ-carrageenan for dysphagic people. Food Res Int [Internet]. 2023 Aug 1 [cited 2024 Sep 15];170:113010. Available from: https://linkinghub.elsevier.com/retrieve/pii/S0963996923005550.
124. Chen Y, McClements DJ, Peng X, Chen L, Xu Z, Meng M, et al. Starch as edible ink in 3D printing for food applications: a review. Crit Rev Food Sci Nutr [Internet] 2024 [cited 2024 Sep 17];64(2):456–471. Available from: https://www.tandfonline.com/action/journalInformation?journalCode=bfsn20.
125. Huang J, Zhang M, Mujumdar AS, Wang Y, Li C. Improvement of 3D printing age-friendly brown rice food on rough texture, swallowability, and in vitro digestibility using fermentation properties of different probiotics. Food Chem [Internet]. 2024 Dec 1 [cited 2024 Sep 15];460:140701. Available from: https://linkinghub.elsevier.com/retrieve/pii/S0308814624023513.

126. Joshi S, Sahu JK, Prakash S, Naik SN. Modulating the 3D printability of vitamin D3-nanoemulsion-based dairy gels: Influence of emulsifier on gel structure, printing behaviour and vitamin D3 retention. J Food Eng [Internet]. 2024 Jul 1 [cited 2024 Sep 15];373:112032. Available from: https://linkinghub.elsevier.com/retrieve/pii/S0260877424000980.
127. Liu Y, Chen K, Zeng Q, Wang P, Zhang Y. The impact of dietary fibers on the construction and molecular network of extrusion-based 3D-printed chicken noodles: unlocking the potential of specialized functional food. Food Chem [Internet]. 2025 Jan 15 [cited 2024 Sep 15];463:141065. Available from: https://linkinghub.elsevier.com/retrieve/pii/S0308814624027158.
128. Liu Y, Zhang Y, Cai L, Zeng Q, Wang P. Protein and protein-polysaccharide composites-based 3D printing: the properties, roles and opportunities in future functional foods. Int J Biol Macromol [Internet]. 2024 Jun 1 [cited 2024 Sep 13];272:132884. Available from: https://linkinghub.elsevier.com/retrieve/pii/S0141813024036894.
129. Kadival A, Mitra J, Machavaram R, Kaushal M. 3D printing of rice starch incorporated peanut protein isolate paste: Rheological characterization and simulation of flow properties. Innov Food Sci Emerg Technol [Internet]. 2024 Jun 1 [cited 2024 Sep 13];94:103669. Available from: https://linkinghub.elsevier.com/retrieve/pii/S1466856424001085.
130. Markovinović AB, Brdar D, Putnik P, Bosiljkov T, Durgo K, Huđek Turković A, et al. Strawberry tree fruits (Arbutus unedo L.): bioactive composition, cellular antioxidant activity, and 3D printing of functional foods. Food Chem [Internet]. 2024 Feb 1 [cited 2024 Sep 15];433:137287. Available from: https://linkinghub.elsevier.com/retrieve/pii/S0308814623019052.
131. Fan M, Choi YJ, Wedamulla NE, Kim SH, Bae SM, Yang D, et al. Different particle sizes of Momordica charantia leaf powder modify the rheological and textural properties of corn starch-based 3D food printing ink. Heliyon [Internet]. 2024;10(4):e24915. Available from: https://linkinghub.elsevier.com/retrieve/pii/S2405844024009460.
132. García-Guzmán L, Cabrera-Barjas G, Soria-Hernández CG, Castaño J, Guadarrama-Lezama AY, Rodríguez LS. Progress in starch-based materials for food packaging applications. In: Polysaccharides, vol. 3. Multidisciplinary Digital Publishing Institute (MDPI); 2022. p. 136–77.
133. Leaw ZE, Kong I, Pui LP. 3D printed corn starch–gelatin film with glycerol and hawthorn berry (Crataegus pinnatifida) extract. J Food Process Preserv. 2021;45(9):e15752.
134. Dermol Š, Borin B, Gregor-Svetec D, Slemenik Perše L, Lavrič G. The development of a bacterial nanocellulose/cationic starch hydrogel for the production of sustainable 3D-printed packaging foils. Polymers (Basel). 2024;16(11):1527.
135. Li S, Jiang Y, Zhou Y, Li R, Jiang Y, Alomgir Hossen M, et al. Facile fabrication of sandwich-like anthocyanin/chitosan/lemongrass essential oil films via 3D printing for intelligent evaluation of pork freshness. Food Chem. 2022;370:131082.
136. Daza LD, Montealegre MÁ, Reche C, Sandoval-Aldana A, Eim VS, Váquiro HA. Chachafruto starch: physicochemical characterization, film-forming properties, and 3D printability. Int J Biol Macromol. 2023;247:125795.
137. Le-Bail A, Maniglia BC, Le-Bail P. Recent advances and future perspective in additive manufacturing of foods based on 3D printing. Curr Opin Food Sci [Internet]. 2020;35:54–64. Available from: https://linkinghub.elsevier.com/retrieve/pii/S2214799320300096.

Chapter 8
Enzymatic Modification of Starch

J. Figueroa-Flórez, L. Hernández-Vanegas, T. Serna-Fadul, and J. Salcedo-Mendoza

8.1 Introduction

Starch is a reserve polysaccharide in plants and a primary source of carbohydrates in the human diet. It is predominantly found in tubers, roots, cereals, seeds, and certain fruits. Starch is synthesized in granular form within plastids: temporarily in chloroplasts and more permanently in amyloplasts. The enzymes starch synthase, branching enzyme, and debranching enzyme regulate the organization of its glucose chains by adding, branching, or removing linkages, ensuring a proper structure and branching pattern [1, 2]. Structurally, starch is composed of two macromolecules: amylose, a linear polymer formed by α-(1, 4) glycosidic linkages, and amylopectin, a mixed-structure polymer with short linear chains that branch via α-(1, 6) glycosidic linkages. The proportion of these components varies depending on the botanical source of the starch, influencing its semicrystalline architecture and governing its techno-functional properties, such as water retention capacity, viscosity, and thermal stability [3, 4].

Native starch granules exhibit technological limitations that restrict their use in the food, pharmaceutical, and bioplastics industries [5]. One of the main challenges is their low stability in water, which leads to retrogradation and gelatinization phenomena, limiting their application under conditions of high temperature and

J. Figueroa-Flórez (✉)
Facultad de Ingeniería, Programa de Ingeniería Agroindustrial, Universidad de Sucre, Sincelejo, Colombia

Facultad de Ingeniería y Administración, Programa de Ingeniería Agroindustrial, Universidad Nacional de Colombia, Palmira, Colombia
e-mail: jorge.figueroa@unisucre.edu.co

L. Hernández-Vanegas · T. Serna-Fadul · J. Salcedo-Mendoza
Facultad de Ingeniería, Programa de Ingeniería Agroindustrial, Universidad de Sucre, Sincelejo, Colombia

G. Ayala Valencia (ed.), *Starch: Progress in Food Applications*,
https://doi.org/10.1007/978-3-031-98732-8_8

humidity [6]. Additionally, their semicrystalline structure contributes to low solubility in cold water and limited capacity to blend uniformly with other materials [7, 8]. Another challenge is starch's low mechanical strength. In the bioplastics industry, starch-based materials tend to be brittle and lack the resilience required for products needing elasticity and flexibility [9]. Moreover, starch exhibits variability in its composition and structure, even when derived from the same botanical source, affecting its gelatinization, pasting, and digestibility properties [10]. These technological limitations and lack of uniformity complicate process standardization and the consistency of final products [11, 12]. These challenges have driven the development of chemical, physical, and enzymatic modification processes to enhance starch functionality and expand its versatility.

Currently, the production of starches modified through physical and enzymatic pathways has gained significant relevance due to restrictions on the use of chemical agents, driven by the high demand for safe and non-toxic products [11, 13]. In this context, enzymatic modification is of great interest because of its specificity and its ability to induce molecular-level changes without the need for harsh chemical reagents, making it particularly relevant for food and pharmaceutical applications [5, 14]. Enzymatic action involves the use of proteins with biocatalytic activity to cleave, reorganize, or structurally modify the starchy macromolecules that compose starch. These proteins catalyze reactions such as hydrolysis, transglycosylation, branching, and debranching [15]. Amylases, such as α-amylase and β-amylase, hydrolyze the α-(1, 4) glycosidic linkages. Glucosidases, or glucoamylases, hydrolyze both α-(1, 4) and α-(1, 6) glycosidic linkages. Transglycosidases facilitate the cleavage of glucose chains and the subsequent formation of cyclic glucose molecules (cyclodextrins) or densely branched molecules. Meanwhile, debranching polypeptides, such as pullulanase and isoamylase, hydrolyze the α-(1, 6) linkages in amylopectin, generating linear chains with varying degrees of polymerization [15–17].

The enzymatic action on starch promotes the production of granules with diverse functionalities, enhancing their versatility and application in the agro-industrial sector. Enzymatically modified starch granules exhibit improved cold-water solubility and have been widely utilized to enhance the texture, stability, and palatability of products such as sauces, soups, and desserts [5, 11]. Moreover, depolymerization facilitates the formation of granules with microporous structures or cyclic molecules with encapsulating properties [8, 18]. In the pharmaceutical field, enzymatically modified starch is used as an excipient, improving drug release and stability [19]. Additionally, the ability of enzymes to produce resistant starches—a type of dietary fiber beneficial for digestive health—has generated increasing interest in the field of nutrition [17, 20].

Recent advances in enzymatic starch modification focus on the concepts of specificity, versatility, and sustainability. However, the effectiveness of this process depends on various factors influencing the interaction between enzymes and starch, ultimately affecting the properties of the modified product. Among the key factors are the composition and structure of starch, which vary depending on the botanical source. Specifically, enzyme affinity for amylose and amylopectin molecules is determined by granule size and the degree of starch crystallinity [21]. Additionally, reaction conditions—such as pH, temperature, reaction time, and substrate/enzyme concentration—are critical variables [22, 23]. Each enzyme has an optimal range of pH and temperature. Generally, the process is conducted at moderate temperatures, below the gelatinization point, and at neutral or slightly acidic pH. Reaction time is crucial, as prolonged periods can lead to complete hydrolysis, producing short fragments or free glucose [5]. Substrate concentration is also important: high concentrations can saturate the enzyme and reduce its efficiency, while lower ratios allow for quicker modification [24]. Furthermore, the presence of inhibitors, such as metal ions, proteins, lipids, or other carbohydrates, can negatively impact the process by reducing enzymatic activity [25]. Lastly, hydrothermal pretreatments of starch can alter its crystalline structure and facilitate enzymatic action by exposing unstable polymer chains susceptible to enzymatic attack. This promotes the formation of more resistant structures [26].

In this context, the present chapter aims to compile information from recent studies on the enzymatic modification of starch under sub-gelatinization conditions and analyze its effects on the structural, morphological, and techno-functional properties of starch. Additionally, it seeks to present technological perspectives on the advantages and potential applications of enzymatic modification in the food industry.

8.2 Starch Granule Structure and Organization: Enzymatic Susceptibility

Starch is a storage carbohydrate synthesized in granular form within cellular organelles known as plastids, through two mechanisms: in chloroplasts, where it is temporarily stored during photosynthesis, and in amyloplasts, where it accumulates for long-term storage. The organization of starch depends on the activity of enzymes responsible for synthesizing glucose chains. Starch synthase adds glucose units to linear chains, branching enzyme introduces α-(1, 6) linkages at branching points, and debranching enzyme removes unnecessary branches to allow for an organized structure and maintain a regular branching pattern [1]. In its natural state, starch granules appear as small particles whose characteristics—such as morphology, size, chemical composition, molecular structure, and susceptibility to enzymatic degradation—are species-specific [27]. The molecular organization of starch can be described in four levels of ordered structure (Fig. 8.1).

The supramolecular organization of starch can be described in four levels, with the macromolecules amylose and amylopectin constituting the first level. Amylose is a linear polyglucan primarily formed by α-(1, 4) linkages, with slight branching that can exceed 10 units. Its structure and size vary depending on the botanical source, with a degree of polymerization ranging from 324 to 4920, a molar mass of 1×10^5 to 1×10^6 g/mol, and a degree of polymerization between 500 and 6000

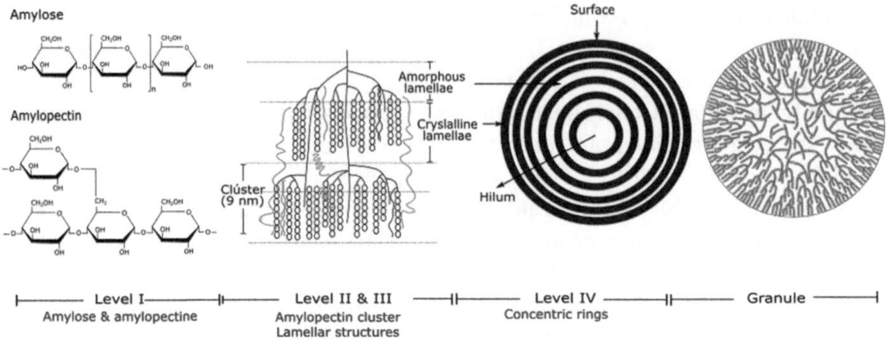

Fig. 8.1 Molecular organization of the starch granule

glucose units [1, 28, 29]. In contrast, amylopectin has a more complex structure due to its high degree of branching, with an average molecular weight of 10^7 to 10^9 Da and a degree of polymerization varying between 10^4 and 10^5 glucose units, depending on the botanical origin of the starch [28, 29]. The second level refers to the amylopectin cluster, which measures 27–28 glucose units, approximately 9 nm. The cluster consists of numerous linear D-glucose chains divided into short chains (A1, B1) and long chains (B2 and B3). Type-A chains (DP 3–12) form double helices, while type-B2 chains (DP 25–36) are present in different clusters and link to other chains via α-(1, 6) bonds. Type-B3 chains (DP >36) can connect several clusters and form the central axis of the amylopectin block. The third level comprises lamellar structures, which are smaller than amylopectin blocks. Crystalline lamellae (3–6 nm) are primarily composed of double helices from the side chains of amylopectin, while amorphous lamellae (4–6 nm) include single helices derived from the branching points of amylopectin. The organization of double helices and their ability to interact with water molecules define the type of semicrystalline polymorph of starch (Fig. 8.2) [29, 30]. The fourth level consists of concentric rings oriented from the hilum toward the granule surface, alternating amorphous and crystalline layers that repeat every 9–10 nm. The thickness of the semicrystalline growth rings gradually decreases, ranging from 450 to 550 nm near the hilum to 80–160 nm in rings closer to the periphery [28, 29]. Finally, at the ultimate level of molecular order, the aggregation and packing of the concentric rings result in the formation of the granule [30, 31].

In general, the biosynthesis of starch is determined by the specific and differentiated biocatalytic action of three enzymes. Consequently, the semicrystalline molecular aggregation of starch into the granule structure can vary depending on its botanical origin. In this context, the intrinsic structure of the granule provides the first level of molecular resistance to enzymatic attack (resistant starch type-II). This means that the granule's susceptibility to enzymatic attacks depends on variables such as the content and location of amylose, the density of branching points, and the presence of pores in the granular structure [32]. These structural characteristics define the diffusion and adsorption processes, which are fundamental mass transfer

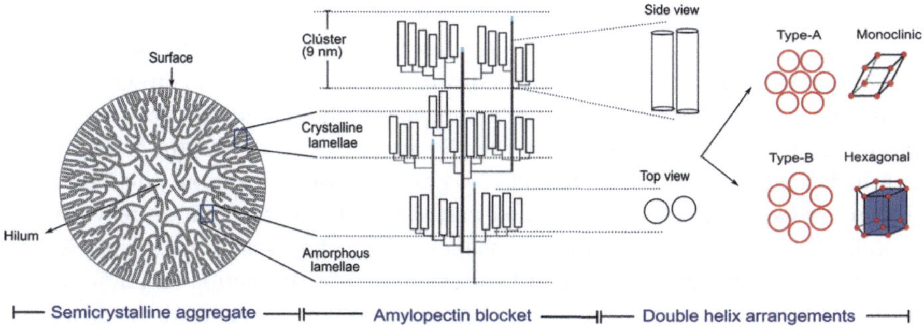

Fig. 8.2 Diffraction patterns and crystal structure of type-A and type-B starches

mechanisms for initiating and activating enzymatic biocatalysis. The presence of innate pores on starch granules can increase susceptibility to enzymatic hydrolysis. According to Pérez et al. [30], the crystalline regions of starch help maintain granule structure, provide stability in the presence of water, and make it more resistant to enzymatic attack. In contrast, the amorphous regions of starch, being less dense, absorb more water at temperatures below the gelatinization point and are therefore more susceptible to enzymatic attack [23]. Other factors, such as the amylopectin-to-amylose ratio, the proportion of short chains, and granule size, are key determinants of the partial or complete hydrolysis of starch [22]. In cereal granules, it has been shown that enzymatic hydrolysis preferentially begins in the amorphous growth rings, causing cracks and lacerations on the granular surface (exoerosion). Additionally, the inherent micro- and nanoporous structure of the granule influences the enzyme's accessibility to the substrate (endoerosion) [5]. In contrast, enzymatic action in tuber starches shows variable results, with surface exoerosion phenomena predominating and no notable presence of internal porosities [1, 5, 23].

8.3 Enzymatic Modification

Native starch exhibits technological limitations that restrict its industrial applications, necessitating structural or morphological modifications to enhance its physicochemical, gelatinization, pasting, and *in vitro* digestibility properties [33]. Modified starch is defined as a derivative of native starch that undergoes changes in one or more properties at the structural, morphological, or physicochemical levels [5]. Starch modification can be achieved through physical, chemical, enzymatic treatments, or a combination of these methods. While chemical modification is effective, it can generate residues that compromise granule safety. In contrast, physical modification is less invasive and does not alter the chemical nature of the granule, directly affecting its molecular organization and size. However, physically modified starches tend to be unstable under extreme conditions of high temperature, humidity, and pH. Enzymatic modification, on the other hand, offers high specificity and greater control over starch structure, making it a promising alternative [5, 11].

Current demands in the food agro-industry call for starch-based products that are cold-water soluble, thermally stable, and have a higher content of resistant starch, produced through environmentally friendly processes. Biocatalytic processes are considered clean technologies that lead to granular disorganization, polymer degradation, molecular rearrangement, and partial or total loss of the granule's native morphology, resulting in significant changes in its techno-functional properties [23, 34]. In this context, enzymatic modification enables the formation of nano-, micro-, or mesoporous structures with increased specific surface area and enhanced absorption capacity. Porous granules improve solubility and liquid absorption, encapsulate bioactive compounds, and slow the release of molecules with nutraceutical properties [23, 35]. Enzymatically hydrolyzed granules also form molecular structures with a higher proportion of short linear chains of low molecular weight, which influence rheological behavior, increase paste clarity, and improve stability during freeze-thaw cycles, preventing syneresis [22, 27, 36]. Moreover, biocatalytic action can reorganize the supramolecular structure of starch by targeting disordered or amorphous regions susceptible to enzymatic action, increasing the fractions of slowly digestible and resistant starch [17, 20]. These modifications expand the application potential of starch in various industries, including food, pharmaceuticals, and biomaterials.

A diverse group of enzymes hydrolyze glycosidic bonds, including hydrolases (α-amylase, β-amylase, amyloglucosidase), debranching enzymes (pullulanase and isoamylase), and transglucosidases (amylosucrase, glucanotransferase, cyclodextrin glycosyltransferase, 4-α-glucanotransferase, amylomaltase, and branching enzyme) [36]. Hydrolases catalyze the cleavage of glycosidic bonds, resulting in the formation of a sugar hemiacetal (glucose) and a free aglycone. Historically, the use of amylases and glucoamylases was limited to the complete hydrolysis of starch. However, over the past two decades, these enzymes have been employed to modify the molecular structure of starch and to investigate the impact of its degradation on granule structure [11, 27]. Additionally, debranching enzymes and transglucosidases have been incorporated for their ability to hydrolyze soluble starch polymers and native starch granules. Debranching enzymes act on the external chains of the amylopectin cluster, producing granules with shorter chains and increased molecular mobility [19, 20, 37]. In contrast, transglucosidases facilitate the transfer of glucan groups from amylose and/or amylopectin (donor) to other molecules (acceptor), leading to the rearrangement of branched chains in amylopectin and the production of cyclic glucans [14–16]. In general, polypeptides with amylolytic activity exhibit varying enzymatic activities depending on their source, extraction method, and purification system. Additionally, factors such as enzyme type, the botanical source of starch, and the ultrastructural characteristics of the granule (chain associations, type and degree of crystallinity, and amylose-lipid interactions) significantly influence the enzymatic hydrolysis process [38, 39].

The hydrolytic action of enzymes exhibits varying levels of starch modification, ranging from simple modifications to dual or more complex processes. Simple processes involve the action of a single enzyme, while complex processes require the biocatalytic activity of two or more enzymes. Multiple modifications can be carried out synergistically in a single stage (simultaneous biocatalysis) or sequentially, where hydrolytic action occurs through individual attacks at different time intervals (sequential biocatalysis) [5, 15]. Studies have shown that combined enzymatic treatments—whether simultaneous or sequential—lead to more significant structural and morphological changes compared to individual enzymatic treatments [11, 15]. These findings highlight the potential of enzyme combinations to modify starch properties, improving its solubility, digestibility, and capacity for microencapsulating bioactive compounds.

8.3.1 Products of Enzymatic Hydrolysis

During the enzymatic modification of starch granules, numerous byproducts known as hydrolysates can be obtained. These are classified based on the number of glucose units they contain (two, three, or more) and the type of polysaccharide structure released (linear, branched, or cyclic). Additionally, the structure of the hydrolysate depends on the enzyme's mechanism of action, whether it acts on the interior or exterior of the starch polymer chains (Fig. 8.3). Hydrolases are subdivided into two types: endoenzymes and exoenzymes.

Endoamylases attack the internal regions of starch, specifically cleaving α-(1, 4) glycosidic bonds in amylose and amylopectin, while excluding α-(1, 6) bonds. This category includes enzymes such as α-amylase (EC 3.2.1.1), maltogenic amylase (EC 3.2.1.133), and malto-oligosaccharide-forming amylases [15]. The hydrolysis products consist of oligosaccharides of varying chain lengths, characterized by the α-configuration at the anomeric carbon (C1) of the reducing glucose unit. According to Benavent and Rosell [40], α-amylase cleaves near the reducing end, preferentially producing maltose, malto-oligosaccharides (DP < 5), and small amounts of α-D-limit dextrins (Fig. 8.4). Maltogenic amylase acts on internal points of amylose molecules, generating free glucose, maltose, maltotriose, and maltopentaose in smaller concentrations, with maltose being the most abundant product of the reaction [41, 42]. Meanwhile, malto-oligosaccharide-forming amylases cleave

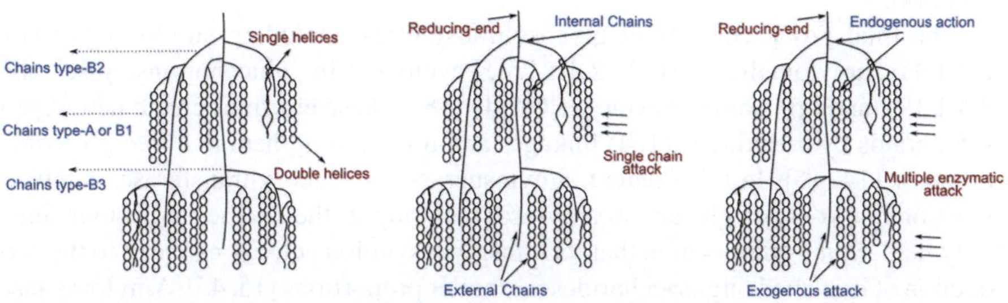

Fig. 8.3 Mode of enzymatic action on starch macromolecules

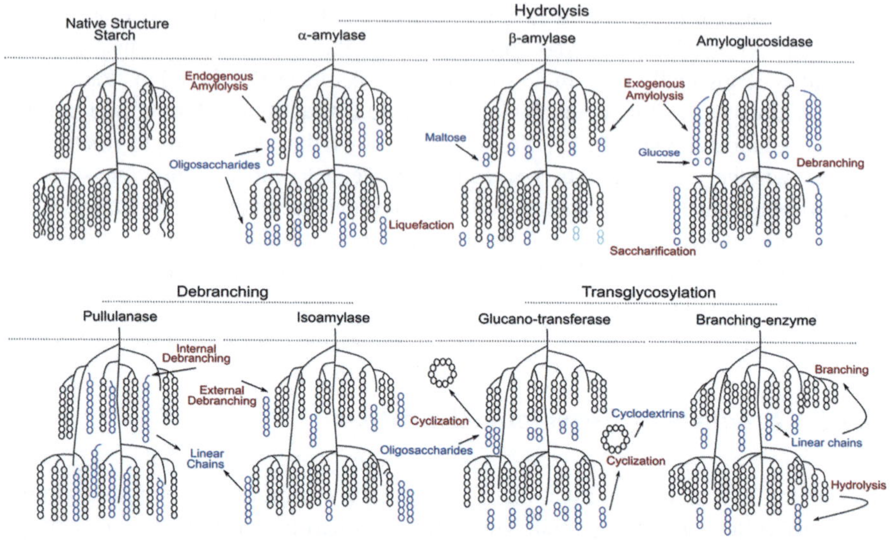

Fig. 8.4 Production of glucose, hydrolysates, and cyclodextrins during starch biocatalysis

glycosidic bonds in amylose and amylopectin, producing small amounts of glucose, maltose, and branched dextrins. These enzymes are characterized by their preferential action in producing malto-oligosaccharides with 2 to 7 glucose units [43]. In contrast, exoamylases, such as amyloglucosidase and β-amylase, are characterized by exogenous action starting from the non-reducing end of the polymeric chain. Amyloglucosidase can simultaneously cleave both α-(1, 4) and α-(1, 6) bonds; thus, its hydrolytic activity results in the production of glucose [15, 41]. β-Amylase preferentially hydrolyzes α-(1, 4) bonds in two-glucose-unit intervals, leading to the production of maltose and β-limit dextrins [15, 21]. Among the hydrolases with debranching activity, pullulanase (EC 3.2.1.41) and isoamylase (EC 3.2.1.68) are notable for their ability to hydrolyze the α-(1, 6) glycosidic bonds in amylopectin [37]. Pullulanase is an endoenzyme that acts within amylopectin, exclusively producing linear polyglycan chains with varying degrees of polymerization, similar to amylose [19]. In contrast, isoamylase performs exogenous action on the branched block, preferentially producing linear chains with a low degree of polymerization [44].

The final group consists of glycosyltransferases, including amylosucrase (EC 2.4.1.4), amylomaltase (EC 2.4.1.25), cyclodextrin glucanotransferase (EC 2.4.1.19), and branching enzymes (EC 2.4.1.18). These enzymes create new glycosidic bonds by breaking α-(1, 4) linkages and transferring them to a free glycosidic acceptor [14, 16]. In this context, amylosucrase, or glucosyltransferase, catalyzes reactions on α-(1, 4) glucans in amylose, resulting in the synthesis of other linear polymers. Some authors note that its transglycosylation activity can lead to the production of branched oligosaccharides in smaller proportions [15, 45]. Amylomaltase, or GTase, cleaves linear α-(1, 4) linkages at an internal point in the chain, releasing a glycosidic segment (dextrin) of variable length. This segment is then transferred and attached to the non-reducing end of another glycosidic chain via a new α-(1, 4) linkage [16, 43]. Cyclodextrin glucanotransferase (CGTase) transforms linear polymers of amylose and/or amylopectin (donors) into other molecules (acceptors), resulting in the rearrangement of amylopectin's branched chains and the production of cyclic glucans.

This enzyme does not perform complete hydrolysis but instead reorganizes α-(1, 4) linkages to form cyclic structures and, to a lesser extent, linear products [14, 27]. Branching enzymes, on the other hand, are glycosyltransferases that catalyze the cleavage of α-(1, 4) linkages in donor chains (7–11 glucose units), releasing a linear glycosidic fragment. This fragment is then transferred to an internal point of the same chain via an α-(1, 6) glycosidic bond, creating a new branch point [15, 27, 43]. In summary, the byproducts of enzymatic hydrolysis are diverse and depend exclusively on two factors: (1) the mechanism of action of the biocatalyst and (2) the position of the glycosidic bond being hydrolyzed in the starch polymer chain.

8.3.2 Effect of Enzymatic Action on Molecular Characteristics

The fundamental properties of the molecular structure of starch include molecular weight, radius of gyration, molecular density, and the distribution of linear chain lengths. Various studies have demonstrated that enzymatic action reduces the molar mass of both polymeric fractions as well as the length of linear chains. This is associated with the cleavage of glycosidic bonds and the release of linear glucans with varying degrees of polymerization [11, 15, 46, 47]. Moreover, results obtained through Size Exclusion Chromatography have identified that the depolymerization of amylose and amylopectin fractions varies depending on the mechanism and mode of action of the enzymes [47–49].

Regarding the action of hydrolases, results obtained through Gel Permeation Chromatography have shown a reduction in amylose and amylopectin proportions in starches treated with bacterial amylases [50, 51]. Similarly, a significant decrease in the radius of gyration and molecular density has been reported for soluble starches hydrolyzed with β-amylase [48]. These findings suggest that longer linear chains are selectively hydrolyzed, forming shorter, less dense, and lower molecular weight polymeric structures. Additionally, analyses performed using High-Performance Anion-Exchange Chromatography have identified an increase in low-degree-of-polymerization linear chains (DP < 12) and a decrease in long chains (DP > 24) following the enzymatic action of amyloglucosidase and β-amylase [48, 50]. Guo et al. [52, 53] also reported a reduction in long chains (DP > 37) over time during the hydrolysis of porous sweet potato starches sequentially treated with maltogenic α-amylase and β-amylase. The reduction in molecular weight and the degree of polymerization of amylopectin linear chains can be attributed to the following factors: (1) The action of amylases on long linear amylose chains, producing relatively shorter chains; (2) the cleavage of internal chains linked by α-(1, 4) bonds within amylopectin clusters, transforming amylopectin into smaller blocks and generating a distribution of shorter linear chains; (3) the leaching of double helices resulting from cleavage at external branch points in amylopectin clusters [47, 48, 52, 53].

Meanwhile, the action of pullulanase reduced the molecular weight of polymeric fractions due to debranching, which hydrolyzes α-(1, 6) glycosidic bonds, releasing a large number of linear chains with a high degree of polymerization [19, 54]. Li et al. [44] examined the effects of modifying rice starch with bacterial pullulanase. The results showed that debranched starches exhibited lower amylose content compared to the native granule. Additionally, the proportion of long-chain amylopectin (> DP 24) decreased with reaction time and the enzymatic activity applied. This was accompanied by a reduction in amylose content and molar mass, indicating the depolymerization of starch molecules into smaller structures with a lower degree of branching. Similarly, the average molar mass decreased in potato starches debranched with pullulanase [54]. Equivalent results were observed in tuber and cereal starches treated with isoamylase [44, 55]. The reduction in molecular weight and the degree of polymerization of amylopectin linear chains, as a result of debranching, is directly associated with the cleavage of α-(1, 6) glycosidic bonds and the consequent loss of linear structures within the amylopectin block [46, 48, 54, 55].

Lastly, the effect of transglycosylation on the molecular properties of enzymatically modified starches has shown similarities to results obtained using hydrolases and debranching enzymes. Lin et al. [56] and Jeong et al. [57] reported a decrease in molar mass and the degree of polymerization of amylopectin linear chains in starches treated with amylomaltase and 4-α-glucanotransferase, depending on the biocatalysis time. This behavior was attributed to the cleavage of linear linkages as well as the rearrangement and reorganization of new linear chains within the hydrolyzed starch granule. It is important to note that transglycosidases, such as 4-α-glucanotransferase and cyclodextrin glucanotransferase, are capable of performing multiple enzymatic actions, including hydrolysis, disproportionation, coupling, and cyclization [14, 41, 58]. Enzymatic modification with branching enzymes involves the formation of shorter chains and new branched structures in starch molecules, leading to significant changes in molecular weight, the degree of chain polymerization, and the proportion of branching points [14, 16, 36]. Treatment of starch with branching enzymes from various microbial sources resulted in an increase in the proportion of α-(1, 6) branch points, a change in the distribution of polymeric structures with a higher presence of short chains, and a reduction in molar mass [59–61]. Starches treated with BE showed a significant proportion of long chains (DP 25–36) and short chains (DP < 6–12), while long chains (DP > 37) and intermediate chains (DP 13–24) decreased compared to native structures [62–64]. Given the catalytic specificity of branching enzymes, it is inferred that the branching reaction includes intermolecular chain transfer to increase the degree of branching and intramolecular reactions in amylopectin to form new branch points. This mode of reaction could explain the observed changes in molecular weight distribution and the new structural configuration of modified starches [63, 64].

Enzymatic modification induces significant changes in the internal structure of starch granules, altering not only their structural components but also their shape

Fig. 8.5 Porous corn granules obtained using amyloglucosidase; cassava granules exo-eroded by amylase; fragmented cassava granules generated with pullulanase; and endo-eroded cassava granules produced using amyloglucosidase. (**a**) Porous (**b**) Exo-erosion (**c**) Fragments (**d**) Endo-erosion. (Reused with permission from Gong et al. [6], Figeroa-Flórez et al. [23, 35], and Acevedo-Viloria et al. [34])

and size, which directly impacts their morphological characteristics (Fig. 8.5) [23]. Generally, the morphological changes observed after enzymatic action depend on the composition, size, and source of the starch granule [11, 40], as well as the type of enzyme, enzyme-to-substrate ratio, and reaction time [11, 15, 65]. In this context, the biocatalytic process, whether individual or simultaneous, has led to the production of starches with porous structures without altering their original morphology [40, 51, 66]. However, enzymatic modification has also produced granules with cracks or lacerations, without clear pore formation [22, 34, 52].

The most influential factors in the morphological changes during the enzymatic hydrolysis of starches are the source of the granule and the type of enzyme used. In cereal starches, hydrolases have caused significant structural changes, resulting in an irregular and microporous surface. These effects include an increase in the number of pores and a rougher granular surface, suggesting greater susceptibility to enzymatic action [49, 51, 67]. On the other hand, the action of α-amylase and amyloglucosidase on tuber starch granules has been less severe, leading to the formation of eroded surfaces and partial loss of the internal polymeric structure, without the formation of pores [22, 23, 35]. This suggests that the action of amylases involves simultaneous exogenous and endogenous activity on the amorphous regions of starch, being more aggressive in smaller granules with micropores. The latter facilitates diffusion and enzymatic absorption processes in the substrate.

However, modifications with pullulanase in tuber starches have resulted in the production of predominantly fragmented granules with slight erosion [53, 54, 67]. This effect has been less pronounced compared to amylases, possibly due to the lower affinity of pullulanase's catalytic sites for starch polymer chains [23, 35]. Additionally, the effect of hydrolysis with transglycosidase enzymes also depends on the botanical source of the starch. In tuber starches, the shape and size of the granules changed significantly after transglycosylation, exhibiting an irregular truncated structure, surface cracks, and slight pore formation [52, 53]. In contrast, in cereal starches, glycosyltransferase enzymes can produce pores of varying sizes, partially disintegrated granules, and the formation of a continuous compact fibrous network with an irregular shape. These results depend on the type of enzyme, reaction time, and temperature [53, 61, 65]. Regarding the type of enzymatic action, it has been observed that the sequential use of amylases with debranching enzymes can cause agglomerations that promote granular compaction and the formation of flattened regions in cassava starch. In contrast, yam granules are slightly affected, showing depressions on the periphery of their typical granular structure [67]. Li et al. [68] reported that the sequential ternary action of hydrolysis, transglycosylation, and debranching in rice starch primarily resulted in porous and irregular structures, with the random presence of more compact fragments exhibiting smooth surfaces. In summary, the results confirm a greater hydrolytic capacity of amylolytic enzymes on cereal granules, regardless of the enzyme or mode of action. Additionally, enzymatic activity shows a preferential attack on the amorphous regions of starch, particularly those located on the outer surface, followed by activity on irregular granules. This leads to the production of fragmented, porous, or slightly eroded starches with a hollow appearance. Furthermore, prolonged hydrolysis times may result in the partial or total loss of the native morphology of the granule.

8.4 Effect of Enzymatic Action on Semicrystalline Order

The semicrystalline order of starch granules determines the behavior of properties such as swelling capacity, gelatinization, pasting, and *in vitro* digestibility. Additionally, the relative crystallinity degree (CD) is closely related to the structure of amylopectin, as double helices significantly contribute to the configuration and arrangement of crystalline lamellae [28, 29]. The effects of enzymatic modification on the diffraction pattern and the relative crystallinity in starches are complex and vary depending on several parameters, including the type of starch, the type of enzyme, the enzyme-to-substrate ratio, the type of enzymatic action (simultaneous or sequential), and the reaction time.

Regarding the type of enzyme, research has demonstrated that enzymatic action of amylases near the initial gelatinization temperature increases the CD of starch without altering the diffraction pattern [21, 23, 49, 51]. Results indicate that amylolysis with α-amylase and β-amylase occurs primarily in the amorphous regions of starch granules, with minimal hydrolysis of the more ordered zones [66, 69]. Salcedo-Mendoza et al. [22] reported an increase in CD without affecting the A-type semicrystalline polymorph in native cassava starch hydrolyzed with α-amylase. Similarly, in wheat and potato starches treated with β-amylase, an increase in CD was observed without changes in the diffraction pattern [21, 70]. The increase in CD is associated with the action of amylolytic enzymes on the amorphous structures of starch, leading to a reduction in the size of amylose linear chains. However, the results are contradictory when glucosidases and debranching enzymes are used. Li et al. [71] revealed that hydrolysis with amyloglucosidase degraded the B-type polymorph (peak 2θ = 5.6) in native yam (*Dioscorea opposita*) starch, suggesting that the enzyme may attack not only the granule's outer layer but also its internal regions. Other studies reported not only a decrease in CD in maize and potato starches treated with pullulanase but also the loss of the native diffraction pattern in sago (*Metroxylon sagu*) starches debranched with different enzyme concentrations [72, 73]. The decrease in CD has been attributed to the hydrolysis of amylopectin branched chains, which are capable of forming double helices and contributing to the crystalline behavior of the granule. In turn, the effect of glycosyltransferases did not alter the CD of modified granules. Some authors suggest that transglycosylation selectively targets the amorphous region and continues with the cleavage of double helices, leading to parallel degradation of the granule's surface and internal layers [12, 74].

In relation to the enzymatic mode of action, the results are varied. The simultaneous action of α-amylase and amyloglucosidase did not produce changes in the diffraction pattern or the relative crystallinity degree of modified cereal and tuber starches. This behavior is likely due to the synergistic action of both enzymes, simultaneously targeting the amorphous and crystalline regions [66, 75]. On the other hand, sequential enzymatic actions have been reported to alter the crystalline order of starch granules. This is attributed to the presence of enzymes capable of hydrolyzing or reorganizing branching points, leading to a new configuration of the molecular density of double helices in amylopectin clusters [52, 53, 64]. With respect to the enzyme-to-substrate ratio, it has been documented that increasing enzyme concentration in the reaction medium causes significant molecular-level changes, drastically affecting starch characteristics, resulting in subsequent alterations in the diffraction pattern and CD [65]. Similar results have been reported with extended enzymatic hydrolysis times. The reduction in the crystalline order of starch granules over time has been attributed to potential layer-by-layer degradation in the following order: first, surface hydrolysis of amorphous layers; followed by an endocorrosion phenomenon in semicrystalline layers; and finally, degradation of crystalline layers. These results reveal the formation of large pores, partial loss of native structure, and a consequent reduction in CD [69, 76].

8.5 Effect of Enzymatic Action on Physicochemical Properties

Solubility and swelling power are key properties that determine the behavior of starch components in aqueous solutions, providing insights into the interactions between the amorphous and crystalline regions of the granule. In general, the amylose and amylopectin content are linked to the solubility and swelling power of starch, respectively [5, 77]. During enzymatic action, the type of enzyme, reaction time, and enzyme-to-substrate ratio are significant factors that promote changes in the constituent macromolecules of starch, directly influencing the granule's ability to hydrate and swell [22, 34, 65, 78, 79]. Previous studies have shown that the use of α-amylase and amyloglucosidase decreases the solubility of starches due to increased crystallinity, a phenomenon attributed to the depolymerization of amylose and the reorganization of linear glucose chains. The reduced solubility in enzymatically modified starches has also been associated with the depolymerization or partial loss of amylose [80, 81]. On the other hand, contradictory results have been reported for starches modified with hydrolases. This phenomenon may be explained by the exo-hydrolytic fragmentation of linear and branched chains, which generates porous, low-molecular-weight polymeric structures that dissolve more readily in water [8, 23, 35, 40, 49]. Additionally, the action of enzymes such as pullulanase has been reported to increase the solubility of starches from various botanical sources. This phenomenon is attributed to the cleavage of branched linkages, reducing the relative crystallinity index and promoting molecular mobility and water solubility mechanisms [46, 67, 82]. Furthermore, enzymatic action by transferases such as transglucosidase has been found to increase solubility in sweet potato granules. This is possibly due to the depolymerization of amylose chains and the consequent formation of small, branched, low-molecular-weight molecules, resulting in a polymeric structure with greater mobility and molecular interaction [40, 52]. The increased solubility of enzymatically modified starches has been attributed to the following factors: (1) leaching of amylose or disruption of the amorphous lamellae, (2) depolymerization and reduction of molecular weight; (3) a higher proportion of short chains; (4) increased availability of hydroxyl groups, and, (5) increased porosity and surface area.

Swelling power has been found to decrease when starch granules are subjected to enzymolysis, as the granular structure weakens and the intermolecular bonds between amylose and amylopectin break, reducing their capacity to retain water and swell [77]. Various studies have reported conflicting and inconsistent results regarding the behavior of swelling power in enzymatically treated starches. The enzymatic action of amyloglucosidase and α-amylase tends to reduce swelling power, which is attributed to the formation of porous structures that weaken the granular structure of starch [23, 49, 70]. However, some enzymatically modified starches have shown an increase in swelling capacity, related to factors such as amylose content, the dense packing of the amylopectin block, and granule size [19]. On the other hand, debranching with pullulanase has led to a decrease in swelling power, attributed to structural changes in the crystalline part of the starch granule. Pullulanase hydrolyzes the α-(1, 6) bonds in branch points, reducing the complexity and packing of the branched structure and increasing the proportion of intermediate linear chains (DP > 13) [72, 82]. Although the action of glycosyltransferases on starch in terms of depolymerization and modification of the branched structure has been documented, there is insufficient research specifically exploring their impact on swelling power [58]. It is therefore crucial to analyze the effect of these enzymes on the reorganization of the amylopectin block and its capacity to hydrate and swell during heating. In summary, the decrease in swelling power can be attributed to: (1) inhibition due to increased amylose content resulting from debranching y, (2) rearrangement of short amylopectin chains and the formation of double helices as a result of transglycosylation. Conversely, an increase in swelling power has been associated with hydrolytic phenomena leading to: (1) leaching and reduction of amylose content, (2) formation of porous structures and, (3) depolymerization of amorphous regions.

8.6 Effect of Enzymatic Action on Pasting and Gelatinization Properties

Gelatinization is a complex process in which starch granules swell and disintegrate in the presence of heat and water, forming a viscous paste or gel. This phenomenon involves the disruption of the ordered amylose structure and the crystalline arrangement as the starch is heated under conditions of sufficient moisture [11, 14]. Gelatinization can be assessed through pasting and gelatinization properties. Experimentally, pasting properties fluctuate depending on time and temperature during consecutive heating and cooling cycles. In this context, a Rapid Visco Analyzer (RVA) is used as an effective tool to measure properties such as peak viscosity, maximum viscosity, breakdown viscosity, and setback viscosity [77]. Additionally, it is important to note that variables affecting the pasting and gelatinization capacity of starch include the amylose-to-amylopectin ratio, chain length distribution, and granule size. In other words, the molecular structure is the key factor determining the viscosities of starch paste or gel [14, 83].

In enzymatically modified starches, the behavior of pasting properties depends on the type of enzyme, hydrolysis time, and the mechanism of action used [15]. Amylolytic enzymes such as α-amylase and β-amylase tend to reduce the viscosities of starch paste and gel, a phenomenon attributed to the hydrolysis of α-(1, 4) glycosidic bonds in amylose chains. This process fragments the long polysaccharide chains into smaller molecules, forming low-molecular-weight structures and reducing the starches' ability to retain water and contribute to granular swelling [34, 40, 55, 84]. Figueroa-Flórez et al. [34], when employing amyloglucosidase on tuber granules, observed similar results in peak viscosity and breakdown viscosity, with a slight decrease in setback viscosity. This outcome is related to the preferential action of amyloglucosidase on the external regions of amylose, compared to the endogenous action of α-amylase, leading to the formation of modified granules with amylose molecules of higher degrees of polymerization. Other studies have demonstrated that the action of debranching enzymes significantly decreases the pasting properties of modified starches [19, 55, 72, 82]. These results may be attributed to the cleavage of branched linkages and the loss of double helices in the amylopectin block, reducing swelling capacity during the heating stage. Regarding the use of glycosyltransferases, a reduction in breakdown and setback viscosities has been reported, associated with alterations in the length and molecular weight of amylose chains [63, 65, 85, 86]. Specifically, maltogenic amylase and cyclodextrin glycosyltransferase have been shown to increase the proportion of short chains in the amylopectin cluster, reducing swelling capacity and peak viscosity during gelatinization. Authors argue that transglycosylation promotes the formation of a highly branched polymeric structure with low amylose content, decreasing pasting properties, particularly mitigating retrogradation, which is reflected in reduced setback viscosity (Table 8.1) [15, 63, 85]. Consequently, the reduction in pasting properties during enzymatic modification of the granule can be associated with various factors, including: (1) The cleavage of double helices; (2) An increase in apparent amylose content; (3) A decrease in the proportion of branches within the amylopectin block [11, 15].

Differential scanning calorimetry (DSC) is the most widely used technique to evaluate the temperature behavior (onset, peak, and conclusion) and the enthalpy of gelatinization. This technique has been employed to demonstrate that enzymatic modification is an effective method to inhibit retrogradation by altering the molecular structure and the rearrangement capacity of the linear and branched starch chains [11, 77]. The literature shows that enzymatic modification increases gelatinization temperatures but decreases gelatinization enthalpy. However, these changes are largely influenced by the type of enzyme and its mechanism of action [11, 15]. For amylases, an increase in gelatinization temperatures has been reported, associated with an increase in the relative crystallinity. This occurs because during amylolysis, enzymes preferentially target the linear regions of amylose, increasing the density of double helices responsible for the crystalline packing in the granule. This molecular packing impedes heat and mass transfer, requiring higher temperatures to disrupt the polymeric chains [23, 35]. On the other hand, debranching induces the loss of double helices and a reduction in the molecular density of the amylopectin block, negatively affecting gelatinization properties [55, 72]. In contrast, the use of glycosyltransferases increases the conclusion temperature and gelatinization enthalpy, improving the thermal stability of starch granules during high-temperature processing [87–89]. This behavior has been associated with the formation of a polymeric structure with a higher proportion of internal amylopectin chains, leading to more flexible branched starch double helices and the formation of higher-quality crystalline clusters.

Gelatinization is closely related to the disruption of amylopectin double helices during the endothermic transition from an ordered semicrystalline structure to an amorphous structure [72]. Thus, variations in enthalpy reflect the loss of double

Table 8.1 Effect of enzyme treatment on functional properties of starch

Enzyme	Starch Source	Effects	Functionality	Reference
α-amylase	Corn	↑ Oil adsorption capacity, Water adsorption capacity ↓ Pasting viscosities, gelatinization enthalpy	Microencapsulation: Modification improved thermal stability of granules, increased specific surface area of porous starch, high resistance to retrogradation	[51]
Amyloglucosidase	Rice	↑ Adsorption capacity, pore volume ↓ Gelatinization temperatures, granule size, crystallinity degree	Results indicate microporous starches could be useful as thermostable adsorbents in food and chemical applications	[66]
β-amylase	Wheat	↑ Amylopectin content, stability during heating, gelatinization temperatures. ↓ Pasting viscosities, crystallinity degree, retrogradation crystals	Gel hardness and retrogradation tendency decreased, suggesting its use in food matrices requiring soft, stable gels during storage	[84]
Pullulanase	Corn, Wheat	↑ Proportion of short chains, water retention capacity, solubility. ↓ Pasting viscosities, granule size	Solubility and water retention analyses indicated that debranched starches are suitable for water capture and hydrogel formation	[46]
Cyclodextrin glycosyltransferase	Corn	↑ Granule size and surface porosity, stability during heating ↓ Peak viscosity, gelatinization enthalpy	CGTase modification at sub-gelatinization temperatures offers an attractive alternative for obtaining porous starches with excellent microencapsulation properties	[85]
Amylosucrase	Corn	↑ Gelatinization temperature, proportion of short chains ↓ Pasting viscosities, crystallinity degree	Modified starches were classified as functional ingredients due to increased resistant starch (RS) and slowly digestible starch (SDS) fractions	[86]

(continued)

Table 8.1 (continued)

Enzyme	Starch Source	Effects	Functionality	Reference
1,4-α-glucan branching enzyme	Cassava	↑ Storage and loss moduli, stability during heating, greater resistance to granule rupture ↓ Crystallinity degree, retrogradation crystals	Gels with higher resistance to rupture, greater stiffness, and reduced retrogradation tendency are ideal for designing starch-based products with specific rheological properties	[83]

helices during enzymatic processes. In this context, results on the gelatinization enthalpy behavior of starch under enzymatic action are varied and sometimes contradictory, primarily depending on the enzymatic mode of action employed. Despite the discrepancies reported in the literature, the main changes in gelatinization enthalpy due to enzymatic effects have been attributed to: (1) variations in polymeric fractions, with a decrease in amylose content and an increase in crystallinity achieved through the use of hydrolases, specifically α-amylase and β-amylase [50, 66, 84, 90]; (2) simultaneous hydrolysis of amorphous and crystalline lamellae by exogenous enzymes such as amyloglucosidase [34, 65, 66, 75]; (3) loss of amylopectin double helices during debranching processes [46, 55, 72, 82]; (3) increased branching density within the amylopectin unit block resulting from the action of branching enzymes [83, 86, 87].

8.7 Effect of Enzymatic Action on *In Vitro* Digestibility Properties

From a nutritional perspective, starch is generally classified into rapidly digestible starch (RDS), slowly digestible starch (SDS), and resistant starch (RS) based on the rate and extent of digestion. RDS causes a rapid increase in blood glucose levels immediately after ingestion, while SDS is fully digested in the small intestine, albeit more slowly. RS, on the other hand, is not digested in the small intestine but is fermented in the large intestine, producing short-chain fatty acids [13, 20]. Studies have shown that the production of SDS and RS through enzymatic processes applied to pregelatinized or retrograded starch granules has been extensively investigated [13, 17, 20]. However, information on the implementation of biocatalytic processes to increase SDS and RS fractions in starch granules under sub-gelatinization conditions is limited.

The literature presents diverse and contradictory results regarding the effect of enzymatic action on *in vitro* digestibility properties, with scientific observations varying specifically based on the botanical origin of the starch. Some studies show an increase in the digestibility rate of modified starches derived from cereals, attributed to the presence of surface pores and internal channels that facilitate the access of digestive enzymes. However, the dense packing of amylose and amylopectin macromolecules in modified tuber granules highlights their molecular resistance to in vitro digestibility [91]. Nonetheless, it is noted that biocatalytic action during enzymatic modification preferentially hydrolyzes amorphous and unstable polymeric materials, which reduces the amount of free glucose accessible to digestive enzymes [66, 69]. Additionally, hydrolases, with their preferential activity on amorphous lamellae, lead to a reduction in the content of rapidly digestible starch (RDS) [20, 91]. These contradictory results highlight that the increase in slowly digestible and resistant starch fractions in enzymatically modified starches is possibly related to the reduction in long-chain lengths (DP > 37), a lower amylose-to-amylopectin ratio, and the formation of shorter polymeric structures [17, 20, 48, 92]. In general, several enzyme groups are used to modify starch granules and produce resistant starch. The results vary depending on the enzymatic mechanism of action and the mode of attack established during experimentation [20]. The use of amylases increased the density of α-(1, 6) glycosidic bonds and the length of short chains (DP < 25), factors that promoted the increase of slowly digestible starch (SDS) and resistant starch (RS) fractions [48, 68]. Debranching action led to modified granules with a higher proportion of short chains (DP < 12) and increased amylose content, which reduced the rapidly digestible starch (RDS) fractions. The findings suggest that the presence of amylose and the outer chains of amylopectin, resulting from the cleavage of α-(1, 6) glycosidic bonds, play a key role in the slow digestion properties. This is because these components, by gaining mobility and interacting with each other or with adjacent amylopectin chains (DP < 24), maximize the SDS and RS content [37, 46, 82]. Additionally, the transglycosylation mechanism increases the density of α-(1, 6) glycosidic bonds in modified starches, exhibiting higher SDS and RS fractions. Authors suggest that the increase in branching points and short chains in the amylopectin block enhances intermolecular network connectivity with amylose, creating more resistant polymeric structures [68, 86]. In this regard, Li et al. (2019) proposed a sequential enzymatic modification with β-amylase, transglucosidase, and pullulanase. The authors reported a significant decrease in the RDS fraction and a notable increase in the RS fraction (over 30%). They argued that the increased relative crystallinity and a higher proportion of short chains (DP < 12) improved molecular mobility, allowing an orderly and stable reorganization of short chains to form new double helices. This resulted in a more ordered and crystalline structure resistant to enzymatic action. In summary, the increase in starch fractions with more resistant digestion behaviors can be attributed to: (1) Ordered and compact lamellar structures; (2) Greater semicrystalline order in starch granules; (3) Thicker crystalline lamellae; (4) A higher proportion of double helices; (5) Increased branching points; (6) A higher proportion of short chains in the amylopectin cluster [20, 93].

8.8 Conclusions

Starch is a highly versatile and cost-low polymer with properties suitable for various specific industrial applications, enhancing its added value despite the known technological limitations in the industry. Recent advancements in starch modification have focused on the concepts of specificity, versatility, safety, and sustainability. In this context, enzymatic modification has demonstrated notable progress, enabling the production of modified granules with diverse technofunctional properties aligned with current market demands, which prioritize the consumption of safe products through clean and eco-friendly processes. Enzymatic action induces significant structural and molecular changes, resulting in granules with enhanced technofunctional properties, such as increased solubility in cold water, reduced gelatinization enthalpy, greater encapsulation capacity, the formation of thermoreversible gels, and a decrease in rapidly digestible starch fractions. However, despite significant progress in the research and development of enzymatically modified starches, scientific challenges remain in designing new functionalities adapted to the technological and economic capabilities of industries. Therefore, it is essential to understand the mechanisms of enzymatic biocatalysis, enzyme substrate specificities, and the intrinsic relationship between the modified structure and starch functionalities. Finally, current scientific knowledge should focus on the development of multi-specific enzymes capable of performing multiple actions in a single step, eliminating the need for simultaneous or sequential biocatalytic processes, to generate complex starch structures in less time and with greater economic viability.

Acknowledgments The authors acknowledge the financial support of the Ministerio de Ciencia, Tecnología e Innovación (MINCIENCIAS) through the National Postdoctoral Fellowship Program No. 924/2023 and Research Project BPIN 2024000100009.

Conflicts of Interest The authors declare no conflict of interest.

References

1. Tappiban P, Smith DR, Triwitayakorn K, Bao J. Recent understanding of starch biosynthesis in cassava for quality improvement: a review. Trends Food Sci Technol. 2019;83:167–80.
2. Xu C, Li C, Li E, Gilbert RG. Insights into wheat-starch biosynthesis from two-dimensional macromolecular structure. Carbohydr Polym. 2024;337:122190.
3. Zhong Y, Li Z, Qu J, Bertoft E, Li M, Zhu F, Blennow A, Liu X. Relationship between molecular structure and lamellar and crystalline structure of rice starch. Carbohydr Polym. 2021;258:117616.
4. Ahmad D, Ying Y, Bao J. Understanding starch biosynthesis in potatoes for metabolic engineering to improve starch quality: a detailed review. Carbohydr Polym. 2024;346:122592.
5. Rostamabadi H, Yildirim-Yalcin M, Demirkesen I, Toker OS, Colussi R, do Nascimento LA, Sahin S, Falsafi SR. Improving physicochemical and nutritional attributes of rice starch through green modification techniques. Food Chem. 2024;458:140212.

6. Gong Y, Xiao S, Yao Z, Deng H, Chen X, Yang T. Factors and modification techniques enhancing starch gel structure and their applications in foods: a review. Food Chemistry: X. 2024;24:102045.
7. Majzoobi M, Farahnaky A. Granular cold-water swelling starch; properties, preparation and applications, a review. Food Hydrocoll. 2021;111:106393.
8. Chen Y, Dai G, Gao Q. Preparation and properties of granular cold-water-soluble porous starch. Int J Biol Macromol. 2020;144:656–62.
9. Yao Y, Hsu Y-I, Uyama H. Enhanced physical properties and water resistance of films using modified starch blend with poly (vinyl alcohol). Ind Crop Prod. 2024;222:119870.
10. Daudt RM, Külkamp-Guerreiro IC, Cladera-Olivera F, Thys RCS, Marczak LDF. Determination of properties of pinhão starch: analysis of its applicability as pharmaceutical excipient. Ind Crop Prod. 2014;52:420–9.
11. Punia-Bangar S, Ashogbon AO, Singh A, Chaudhary V, Whiteside WS. Enzymatic modification of starch: a green approach for starch applications. Carbohydr Polym. 2022;287:119265.
12. Wu K, Li C, Li Z, Gu Z, Ban X, Hong Y, Cheng L, Kong H. Enzymatic modification lowers syneresis in corn starch gels during freeze-thaw cycles through 1,4-α-glucan branching enzyme. Int J Biol Macromol. 2024;269:132183.
13. Zhang Z, Bao J. Recent advances in modification approaches, health benefits, and food applications of resistant starch. Starke. 2021;75:2100141.
14. Wu K, Li C, Li Z, Li Z, Gu Z, Ban X, Hong Y, Cheng L, Kong H. Highly-branched modification of starch: an enzymatic approach to regulating its properties. Food Hydrocoll. 2024;147:109433.
15. Negi A, Barthwal R, Kathuria D, Singh N. Enzymatic advances in starch modification: creating functional derivatives and exploring applications. Food Biosci. 2024;62:105074.
16. Wang Y, Møller MS, Svensson B. Transglycosylase structures and their role in granular starch modification: a mini review. Food Biosci. 2024;59:104095.
17. Si H, Xie F, Yang R, Gu W, Wu S, Zhang J, Zhang Y, Qiao Y. Recent developments in enzymatic preparation, physicochemical properties, bioactivity, and application of resistant starch type III from staple food grains. Int J Biol Macromol. 2024;279:135521.
18. Jeong HM, Kang HN, Kim M-Y, Cho HS, Baek J-W, Yu YJ, Shim JH. Enhanced physicochemical properties of isoflavone derivatives through enzymatic encapsulation using cyclodextrin glucanotransferase. Food Hydrocoll. 2024;23:110874.
19. Liu G, Hong Y, Gu Z, Li Z, Cheng L, Li C. Preparation and characterization of pullulanase debranched starches and their properties for drug controlled-release. RSC Adv. 2015;5(117):97066–75.
20. Ismail A, Saradhuldhat W, Tananuwong K, Krusong K. Enzymes for resistant starch production. Food Biosci. 2024;16:105529.
21. Das R, Kayastha AM. Enzymatic hydrolysis of native granular starches by a new β-amylase from peanut (*Arachis hypogaea*). Food Chem. 2019;276:583–90.
22. Salcedo-Mendoza J, Paternina-Urzola S, Lujan-Rhenals D, Figueroa-Flórez J. Enzymatic modification of cassava starch (Corpoica M-Tai) around the pasting temperature. Dyna. 2018;85(204):223–30.
23. Figueroa-Flórez J, Edith Cadena-Chamorro E, Rodríguez-Sandoval E, Salcedo-Mendoza J, Ciro-Velásquez H. Hydrothermal processes and simultaneous enzymatic hydrolysis in the production of modified cassava starches with porous-surfaces. Heliyon. 2023;9(7):e17742.
24. Kong H, Yang X, Gu Z, Li Z, Cheng L, Hong Y, Li C. Heat pretreatment improves the enzymatic hydrolysis of granular corn starch at high concentration. Process Biochem. 2018;64:193–9.
25. Ai Y, Hasjim J, Jane JL. Effects of lipids on enzymatic hydrolysis and physical properties of starch. Carbohydr Polym. 2013;92(1):120–7.
26. Almeida RLJ, Santos NC, de Brito Lima WB, de Araújo Padilha CE, Rios NS, Dos Santos ES. Effect of enzymatic hydrolysis on digestibility and morpho-structural properties of hydrothermally pre-treated red rice starch. Int J Biol Macromol. 2022;222(A):65–76.

27. Park S, Kim YR. Clean label starch: production, physicochemical characteristics, and industrial applications. Food Sci Biotechnol. 2021;30(1):1–17.
28. Vamadevan V, Liu Q. Starch: starch architecture and structure. In: Encyclopedia of food grains. Elsevier Science & Technology; 2016. p. 190–7.
29. Bertoft E. Analyzing starch molecular structure. In: Starch in food. Elsevier Science & Technology; 2018. p. 97–149.
30. Pérez S, Baldwin P, Gallant D. Structural features of starch granules I. In: Starch: chemistry and technology. Academic Press; 2012. p. 149–92.
31. Wang S, Xu H, Luan H. Multiscale structures of starch granules. In: Starch structure, functionality and application in foods. Springer; 2020. p. 41–55.
32. Wang S, Guo P. Botanical sources of starch. In: Starch structure, functionality and application in foods. Springer; 2020. p. 9–21.
33. Figueroa-Flórez JA, Cadena-Chamorro EM, Rodríguez-Sandoval E, Salcedo-Mendoza JG, Ciro-Velásquez HJ. Cassava starches modified by enzymatic biocatalysis: effect of reaction time and drying method. Dyna. 2019;86(208):162–70.
34. Acevedo-Viloria N, Cervera-Ricardo M, Figueroa-Flórez J, Salcedo-Mendoza J, Ramos-Villacob V. Vigilancia científica y tecnológica en procesos de modificación físico-enzimática en gránulos de almidón. Revista UDCA Actualiadad & Divulgación Cientifica 2024; 27(1).
35. Figueroa-Flórez JA, Arroyo Dagobeth ED, Cadena-Chamorro E, Rodríguez-Sandoval E, Salcedo-Mendoza JG, Ciro-Velásquez HJ. Effect of physical and thermal pretreatments on enzymatic activity in the production of microporous cassava starch. Agron Colomb. 2022;41(1):e105089.
36. Zhong Y, Xu J, Liu X, Ding L, Svensson B, Herburger K, Guo K, Pang C, Blennow A. Recent advances in enzyme biotechnology on modifying gelatinized and granular starch. Trends Food Sci Technol (Internet). 2022;123:343–54.
37. Liu G, Gu Z, Hong Y, Cheng L, Li C. Structure, functionality and applications of debranched starch: a review. Trends Food Sci Technol (Internet). 2017;63:70–9.
38. Xia W, Zhang K, Su L, Wu J. Microbial starch debranching enzymes: developments and applications. Biotechnol Adv. 2021;50(107786):107786.
39. Farias TC, Kawaguti HY, Bello Koblitz MG. Microbial amylolytic enzymes in foods: technological importance of the *Bacillus* genus. Biocatal Agric Biotechnol. 2021;35:102054.
40. Benavent-Gil Y, Rosell CM. Morphological and physicochemical characterization of porous starches obtained from different botanical sources and amylolytic enzymes. Int J Biol Macromol. 2017;103:587–95.
41. Sun S, Li R, Sun D, Guo L, Cui B, Zou F. Improving paste stabilities of cassava starch through molecular density after maltogenic amylase and transglucosidase. Food Chem. 2025;462:140993.
42. Liu P, Ma L, Duan W, Gao W, Fang Y, Guo L, Yuan C, Wu Z, Cui B. Maltogenic amylase: its structure, molecular modification, and effects on starch and starch-based products. Carbohydr Polym. 2023;319:121183.
43. Miao M, Jiang B, Jin Z, BeMiller JN. Microbial starch-converting enzymes: recent insights and perspectives: microbial starch-converting enzyme. Compr Rev Food Sci Food Saf. 2018;17(5):1238–60.
44. Li Y, Xu J, Zhang L, Ding Z, Gu Z, Shi G. Investigation of debranching pattern of a thermostable isoamylase and its application for the production of resistant starch. Carbohydr Res. 2017;446–447:93–100.
45. Wang Y, Li X, Ji H, Zheng D, Jin Z, Bai Y, Svensson B. Thermophilic 4-α-glucanotransferase from *Thermoproteus uzoniensis* retards the long-term retrogradation but maintains the short-term gelation strength of tapioca starch. J Agric Food Chem. 2020;68(20):5658–67.
46. Liu G, Hong Y, Gu Z, Li Z, Cheng L. Pullulanase hydrolysis behaviors and hydrogel properties of debranched starches from different sources. Food Hydrocoll. 2015;45:351–60.
47. Wang D, Zheng X, Liu W, Sun Q, Chen H-H, Mu H. Preparation and characterization of debranched starches: influence of botanical source and debranching time. Food Chem. 2023;407:135141.

48. Miao M, Li R, Huang C, Jiang B, Zhang T. Impact of β-amylase degradation on properties of sugary maize soluble starch particles. Food Chem. 2015;177:1–7.
49. Keeratiburana T, Hansen AR, Soontaranon S, Tongta S, Blennow A. Porous rice starch produced by combined ultrasound-assisted ice recrystallization and enzymatic hydrolysis. Int J Biol Macromol. 2020;145:100–7.
50. Villas-Boas F, Franco CML. Effect of bacterial β-amylase and fungal α-amylase on the digestibility and structural characteristics of potato and arrowroot starches. Food Hydrocoll. 2016;52:795–803.
51. Song Z, Zhong Y, Tian W, Zhang C, Hansen AR, Blennow A, Liang W, Guo D. Structural and functional characterizations of α-amylase-treated porous popcorn starch. Food Hydrocoll. 2020;108:105606.
52. Guo L, Li H, Lu L, Zou F, Tao H, Cui B. The role of sequential enzyme treatments on structural and physicochemical properties of cassava starch granules. Starke. 2019;71(7–8):1800258.
53. Guo L, Li J, Li H, Zhu Y, Cui B. The structure property and adsorption capacity of new enzyme-treated potato and sweet potato starches. Int J Biol Macromol. 2020;144:863–73.
54. Asiri SA, Flöter E, Ulbrich M. Enzymatic modification of granular potato starch - effect of debranching on morphological, molecular, and functional properties. Starke. 2019;71(9–10):1900060.
55. Mendez-Montealvo G, Wang YJ, Campbell M. Thermal and rheological properties of granular waxy maize mutant starches after isoamylase modification. Carbohydr Polym. 2011;83(4):2011–5.
56. Lin H-T, Yeh A-I, Ho T-HD, Pan M-H, Chen SH. Molecular weight, chain length distribution and long-term retrogradation of cassava starch modified by amylomaltase. Food Hydrocoll. 2023;134:108027.
57. Jeong DW, Jeong HM, Shin YJ, Woo S-H, Shim JH. Properties of recombinant 4-α-glucanotransferase from Bifidobacterium longum subsp. longum JCM 1217 and its application. Food Sci Biotechnol. 2020;29(5):667–74.
58. Chen Y, McClements DJ, Peng X, Chen L, Xu Z, Meng M, Ji H, Long J, Qiu C, Zhao J, Jin Z. Research progresses on enzymatic modification of starch with 4-α-glucanotransferase. Trends Food Sci Technol. 2023;131:164–74.
59. Dong Y, Li Z, Kong H, Ban X, Gu Z, Zhang H, Hong Y, Chen L, Li C. Correlation analysis of starch molecular structure and film properties via rearrangements of glycosidic linkages by 1,4-α-glucan branching enzyme. Carbohydr Polym. 2025;348(Pt B):122908.
60. Zhu J, Li X, Lu C, Zhou X, Chen L, Qiu C, Jin Z, Lomg J. An efficient approach for the preparation of branched starch through thermophilic glycogen branching enzyme modification. Food Hydrocoll. 2024;156:110339.
61. Wang Y, Pang C, Mohammad-Beigi H, Li X, Wu Y, Lin MK, Bai Y, Moller MS, Svensson B. Sequential starch modification by branching enzyme and 4-α-glucanotransferase improves retention of curcumin in starch-alginate beads. Carbohydr Polym. 2024;323:121387.
62. Gaenssle ALO, van der Maarel MJEC, Jurak E. The influence of amylose content on the modification of starches by glycogen branching enzymes. Food Chem. 2022;393:133294.
63. Xia C, Zhong L, Wang J, Zhang L, Chen X, Ji H, Ma S, Dong W, Ye X, Huang Y, Li Z, Cui Z. Structural and digestion properties of potato starch modified using an efficient starch branching enzyme AqGBE. Int J Biol Macromol. 2021;184:551–7.
64. Zhong Y, Herburger K, Kirkensgaard JJK, Khakimov B, Hansen AR, Blennow A. Sequential maltogenic α-amylase and branching enzyme treatment to modify granular corn starch. Food Hydrocoll. 2021;120:106904.
65. Benavent-Gil Y, Rosell CM. Comparison of porous starches obtained from different enzyme types and levels. Carbohydr Polym. 2017;157:533–40.
66. Lacerda LD, Leite DC, Soares RMD, da Silveira NP. Effects of α-amylase, amyloglucosidase, and their mixture on hierarchical porosity of rice starch. Starke. 2018;70(11–12):1800008.
67. Li H, Li J, Xiao Y, Cui B, Fang Y, Guo L. In vitro digestibility of rice starch granules modified by β-amylase, transglucosidase and pullulanase. Int J Biol Macromol. 2019;136:1228–36.

68. Arroyo-Dagobeth ED. Implementación de procesos biocatalíticos e hidrotérmicos en la modificación de mezclas poliméricas de almidones nativos de yuca y ñame. (Facultad de Ciencias, Escuela de Biociencias): Universidad Nacional de Colombia; 2022. p. 185.
69. Figueroa-Flórez J, Cadena-Chamorro E, Salcedo-Mendoza J, Rodríguez-Sandoval E, Ciro-Velásquez H, Serna-Fadul T. Enzymatic biocatalysis processes on the semicrystalline and morphological order of native cassava starches (*Manihot esculenta*). Rev Fac Nac Agron Medellin. 2024;77(3):10839–52.
70. Li Y, Cheng W, Qiu X, Sun Y, Xia X, Yang L, Fan M, Wang L, Qian H. Effects of β-amylase hydrolysis on the structural, physicochemical and storage properties of wheat starch. J Cereal Sci. 2023;109:103605.
71. Li X, Gao W, Wang Y, Jiang Q, Huang L. Granule structural, crystalline, and thermal changes in native Chinese yam starch after hydrolysis with two different enzymes–α-amylase and gluco-amylase. Starke. 2011;63(2):75–82.
72. Li P, He X, Dhital S, Zhang B, Huang Q. Structural and physicochemical properties of granular starches after treatment with debranching enzyme. Carbohydr Polym. 2017;169:351–6.
73. Leong YH, Karim AA, Norziah MH. Effect of pullulanase debranching of sago (*Metroxylon sagu*) starch at subgelatinization temperature on the yield of resistant starch. Starke. 2007;59(1):21–32.
74. Hassanein WS, İspirli H, Dertli E, Yilmaz MT. Structural characterization of potato starch modified by a 4,6-α-glucanotransferase B from *Lactobacillus reuteri* E81. Int J Biol Macromol. 2023;242(Pt 3):124988.
75. Wasserman LA, Papakhin AA, Borodina ZM, Krivandin AV, Sergeev AI, Tarasov VF. Some physico-chemical and thermodynamic characteristics of maize starches hydrolyzed by glucoamylase. Carbohydr Polym. 2019;212:260–9.
76. Foresti ML, del Williams MP, Martínez-García R, Vázquez A. Analysis of a preferential action of α-amylase from *B. licheniformis* towards amorphous regions of waxy maize starch. Carbohydr Polym. 2014;102:80–7.
77. Salcedo-Mendoza J, Figueroa-Flórez J, Hernández-Ruidíaz J, Cervera-Ricardo M. Agroindustria de materiales amiláceos: Métodos y Protocolos de Caracterización. Universidad de Sucre; 2024. p. 222.
78. Dura A, Błaszczak W, Rosell CM. Functionality of porous starch obtained by amylase or amyloglucosidase treatments. Carbohydr Polym. 2014;101:837–45.
79. Martínez-Alejo JM, Benavent-Gil Y, Rosell CM, Carvajal T, Martínez MM. Quantifying the surface properties of enzymatically-made porous starches by using a surface energy analyzer. Carbohydr Polym. 2018;200:543–51.
80. Shariffa YN, Uthumporn U, Karim AA, Haiyee ZA. Hydrolysis of native and annealed tapioca and sweet potato starches at subgelatinization temperature using a mixture of amylolytic enzymes. Int Food Res J. 2027;24:1925–33.
81. Hutabarat DJC, Stevensen J. Physicochemical properties of enzymatically modified starch: a review. IOP Conf Ser Earth Environ Sci. 2023;1169(1):012093.
82. Ge X, Shen H, Su C, Zhang B, Zhang Q, Jiang H, Yuan L, Yu X, Li. Pullulanase modification of granular sweet potato starch: assistant effect of dielectric barrier discharge plasma on multi-scale structure, physicochemical properties. Carbohydr Polym. 2021;272:118481.
83. Li Y, Li C, Gu Z, Hong Y, Cheng L, Li Z. Effect of modification with 1,4-α-glucan branching enzyme on the rheological properties of cassava starch. Int J Biol Macromol. 2017;103:630–9.
84. Cheng W, Sun Y, Xia X, Yang L, Fan M, Li Y, Wang L, Qian H. Effects of β-amylase treatment conditions on the gelatinization and retrogradation characteristics of wheat starch. Food Hydrocoll. 2022;124:107286.
85. Dura A, Rosell CM. Physico-chemical properties of corn starch modified with cyclodextrin glycosyltransferase. Int J Biol Macromol. 2016;87:466–72.
86. Wang R, Zhang T, He J, Zhang H, Zhou X, Wang T, Feng W, Chen Z. Tailoring digestibility of starches by chain elongation using amylosucrase from neisseria polysaccharea via a zipper reaction mode. J Agric Food Chem. 2020;68(1):225–34.

87. Tian Y, Chen H, Zhang X, Zhan J, Jin Z, Wang J. Highly branched dextrin prepared from high-amylose maize starch using waxy rice branching enzyme (WRBE). Food Chem. 2016;203:530–5.
88. Li H, Dhital S, Slade AJ, Yu W, Gilbert RG, Gidley MJ. Altering starch branching enzymes in wheat generates high-amylose starch with novel molecular structure and functional properties. Food Hydrocoll. 2019;92:51–9.
89. Gui Y, Zou F, Li J, Tang J, Guo L, Cui B. Corn starch modification during endogenous malt amylases: the impact of synergistic hydrolysis time of α-amylase and β-amylase and limit dextrinase. Int J Biol Macromol. 2021;190:819–26.
90. Li H, Cui B, Janaswamy S, Guo L. Structural and functional modifications of kudzu starch modified by branching enzyme. Int J Food Prop. 2019;22(1):952–66.
91. Das M, Santra S, Chakraborty M, Rajan N, Sarvanabhupathy S, Anusha BP, Banerjee R. Resistant starch: insights into better health and metabolism. Biocatal Agric Biotechnol. 2024;59:103275.
92. Shi M, Zhang Z, Yu S, Wang K, Gilbert RG, Gao Q. Pea starch (*Pisum sativum* L.) with slow digestion property produced using β-amylase and transglucosidase. Food Chem. 2014;164:317–23.
93. Gebre BA, Zhang C, Li Z, Sui Z, Corke H. Impact of starch chain length distributions on physicochemical properties and digestibility of starches. Food Chem. 2024;435:137641.

Chapter 9
Starch in Plant-based Foods

Guilherme Dallarmi Sorita, Wilson Daniel Caicedo Chacon, Mikaela Rajchman, and Eduardo Leonarski

9.1 Introduction

Plant-based replacement products have garnered significant global interest, with projections indicating continued growth in the upcoming years. The shift toward plant-based diets presents a complex public health challenge closely tied to environmental sustainability. Encouraging a global move towards nutrient-dense, plant-based diets is one of the most impactful strategies for building sustainable food systems that can nourish the estimated global population of 10 billion people by 2050 [1].

In this context, starch is an important ingredient and a critical component that enhances yield, viscosity, water-holding capacity, gel formation, emulsification, and surface hydrophobicity of protein molecules. Moreover, it adds body, creates fibrous structures, and improves mouthfeel, thereby enhancing the texture and stability of plant-based products to make them more palatable and appealing [2–6]. For example, starch can replicate the fibrous texture of animal proteins in plant-based meat analogs [2, 3] and add creaminess to dairy substitutes [7–10]. This versatility enables food manufacturers to develop innovative products that meet consumer preferences and dietary needs.

G. D. Sorita (✉) · E. Leonarski
Department of Chemical and Food Engineering, Federal University of Santa Catarina, Florianópolis, SC, Brazil

W. D. C. Chacon
Department of Chemical and Food Engineering, Federal University of Santa Catarina, Florianópolis, SC, Brazil

Ingeniería de Alimentos, Fundación Universitaria Agraria de Colombia, Bogotá, Colombia

M. Rajchman
Institute of Food Science Research - CIAL (CSIC-UAM), Madrid, Spain

G. Ayala Valencia (ed.), *Starch: Progress in Food Applications*,
https://doi.org/10.1007/978-3-031-98732-8_9

Starch can be derived from various plant sources, including roots (*e.g.*, cassava and potatoes) [11], tubers (*e.g.*, sweet potatoes) [11], and cereals (*e.g.*, corn, rice, and wheat) [12]. Each source brings unique properties that allow producers to customize formulations to achieve specific desired technological properties.

Starch consists primarily of two polysaccharides, amylose and amylopectin, each with distinct functional properties [12]. Amylose, a linear glucose polymer, contributes to gelation and thickening, while amylopectin, a branched polymer, enhances creaminess and viscosity [12–14]. The amylose-to-amylopectin ratio varies across starch sources, impacting on digestibility, gelatinization, and retrogradation, all qualities that can be optimized to achieve the desired properties in plant-based products [13, 15].

Starch plays a crucial role in maintaining certain attributes of plant-based foods throughout processing, cooking, and storage. It undergoes transformations such as gelatinization, retrogradation, and syneresis, which directly affect texture, consistency, and shelf life [16]. For example, when heated with water, starch granules swell and gelatinize, thickening and stabilizing food matrices [13]. This property is essential for creating desirable textures in plant-based meats, sauces, and dairy alternatives.

However, employing starch in plant-based foods presents challenges, particularly in maintaining stability during cooking and processing, directly impacting product quality and shelf life [2, 5, 6]. Additionally, interactions between starch and other components, such as proteins, fats, and fibers, influence these products' sensory and nutritional profiles [4]. Effective management of these processes, along with sensory studies throughout the whole shelf life, are essential to produce plant-based foods with consistent texture, structure, and freshness that meet consumer expectations.

Based on this, this chapter delves into the functional roles of starch in plant-based food systems, examining how starch and its interactions with other macromolecules affect food properties and nutritional quality. It also explores starch transformations during processing, cooking, and storage, the stability of plant-based products over time, and innovative starch applications. Finally, it addresses sensory characteristics, technological and regulatory challenges, and future research directions for advancing starch utilization in plant-based food development.

9.2 Innovative Uses of Starch in Plant-Based Foods

The application of starch in plant-based analog products has been extensively explored over the last few years. The innovative use of starch in alternative plant-based products, such as meat, dairy, and egg analogs, underscores the key role of this polysaccharide in simulating the functional and sensory characteristics of animal-based foods. Understanding the different conventional functions of starch, how it interacts with other molecules (like proteins and fats), and the processing methods applied (e.g., gelatinization, heating) is essential to drive further

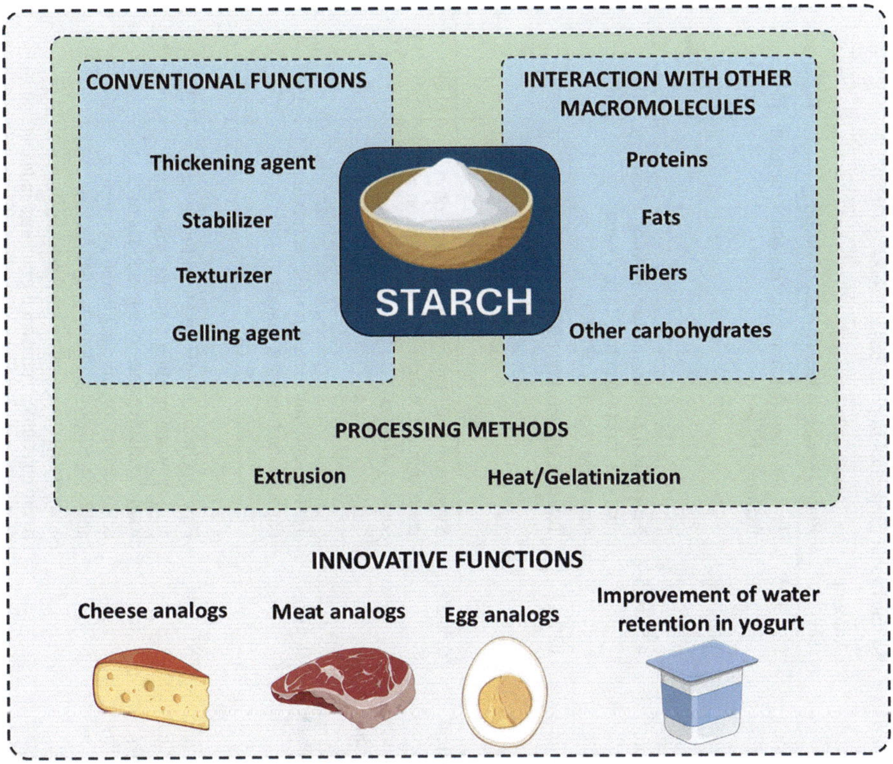

Fig. 9.1 Innovative functions of starch in plant-based foods

innovation in plant-based foods (Fig. 9.1). These interactions and processing techniques not only enhance textural qualities but also stabilize the product's structure, making it possible to develop analogs that closely resemble animal-based textures and sensory profiles. Table 9.1 summarizes the most significant studies on the use of starch in plant-based food analogs.

The popularity of dairy products has led to the creation of alternative plant-based dairy products, which aim to replicate the texture, flavor, and sensory experience of traditional dairy products while using plant-based ingredients. Researchers have explored various ingredients, such as plant proteins (*e.g.*, soy protein, pea protein), fats (*e.g.*, coconut oil, sunflower oil), and starches (*e.g.*, tapioca starch, potato starch), to create a matrix with the desired properties and address this challenge.

Starch can act as a thickener or stabilizer in dairy substitutes, contributing to the creamy texture of plant-based yogurts, cheeses, and ice creams [2, 4, 6, 10, 20, 25]. A typical example of a dairy product analog is cheese. Yun et al. [6] conducted a study to develop a cheese analog formulation and to identify the most suitable starch with melt-stretch properties among nine different types (waxy potato, tapioca, potato, sweet potato, glutinous rice, hydroxypropyl tapioca, acetylated distarch adipate, and α-corn), which are essential for enhancing the functional characteristics of

Table 9.1 Applicability of starch for enhancing functional, technological, sensory, and nutritional properties in plant-based products

Plant-based product	Starch source	Functionality	Amount (%)	Plant-based product characterization	Reference
Cheese analogs	Waxy potato, tapioca, potato, sweet potato, glutinous rice, hydroxypropyl tapioca acetylated distarch adipate, and α-corn starches	Improve melt-stretchability	12.4%	Texture, rheological properties and structural features (FTIR)	[6]
Cheese analogs	Pea, corn, potato, wheat, tapioca, and waxy	Improve the texture and rheological properties	4.5–10.5%	Nutritional profile, texture, rheology, sensory analysis, microstructure (MEV) and structural features (FTIR)	[17]
High-protein plant-based cheese alternative	Waxy corn starch	Enhance the viscoelastic properties	12%	Texture, meltability, oil loss and rheology	[9]
Yogurt-like	Tapioca starch	Improve physicochemical, rheological, and sensory properties, reduce syneresis and improve water holding capacity	0.5–2%	Syneresis, Rheological properties, color measurement, nutritional composition and sensory analysis	[18]
Plant-extract beverages like Greek-style yogurt	Waxy maize starch	Improve texture and sensory properties	0–16%	Texture and nutritional profile, antioxidant capacity and sensory analysis	[19]
Fermented yogurt-like	Quinoa flour	Viscosity increased: nutritional characteristics remain stable or slightly increas, good textural (by sensory analysis) and organoleptic profiles	Ni	Physicochemical (pH and total titratable acidity, water holding capacity, viscosity, total dry matter and color), kinetics of acidification, organic acids and free amino acids profile, total phenols, antioxidant activity, viscosity, nutritional profile, starch hydrolysis index, predicted glycemic index and sensory profile	[20]

Plant-based ice cream	Rice, lentil, and chickpea flours (ratio 4:2:1)	Starch gelatinization improved the creamy-viscous structure and good sensory acceptance	Ni	Biochemical analysis, proximal composition, technological characterization (dry matter, overrun, apparent viscosity, color, and texture) and sensory profile	[10]
Meat analogs	Corn starch extruded with wheat gluten	Reduced cooking loss, improved sensory attributes	10–20%	Texture (solids lost and water-holding capacity) and sensory analysis	[21]
Meat analogs	Cornstarch	Reduced cook loss (%) and good shape retention during cooking and good sensory acceptability	20–40%	Physical changes after cooking (shape retention and total yield through cook loss, fat loss, diameter reduction and increased thickness and texture) and sensory attributes	[21]
Salmon fillet simulants	Cornstarch	The bioinks contain comparable microstructure and similar macronutrient composition to real salmon. Good structural stability, high print resolution, and support for microwave cooking.	Ni	3D-printed salmon fillet simulants, Rheological properties, microstructure (confocal laser scanning microscopy, SEM), structural properties (FTIR), texture, stability to heating	[22]
Extruded meat replacer (based on peanut protein)	Wheat	Improved texture	2.8%	Texture, microstructure (SEM), rheological properties, thermal properties (DSC and TGA) and structural properties (FTIR)	[23]
Egg analogs	Potato, corn, and glutinous rice	Improved the hardness, springiness, cohesiveness, chewiness, and specific volume	14% of soy protein isolate/starch blends	Texture, air holding properties, rheological properties, molecular interaction, structural properties (FTIR and X-ray diffraction), microstructure (MEV), *in vitro* digestion	[4]

(continued)

Table 9.1 (continued)

Plant-based product	Starch source	Functionality	Amount (%)	Plant-based product characterization	Reference
Vegan omelet analog	Potato starch	Reduction of water loss and increasing oil adsorption during frying, altering thermal gelation and improved sensory perception	6.0% (w/w)	Sensory evaluation, physicochemical (water content and total oil absorption), rheological characteristics, microstructure (confocal laser scanning microscopy), *in vitro* digestion, thermal properties (DSC) and structural properties (NMR)	[24]
Plant-based emulsifier for mayonnaise	Modified rice starch—Esterification by Octenyl succinic anhydride (OSA)	High emulsion index, viscosity and storage stability	200 mg starch mL^{-1} oil	Particle size distribution and optical microscopy, Emulsion index, rheological characteristics and stability test	[5]

Ni not informed

plant-based cheese analogs. The authors found that the addition of 12.4% of waxy potato, tapioca, or hydroxypropyl tapioca starches exhibited excellent melt-stretchability in the cheese analog formulation, as confirmed by changes in viscoelastic properties over temperature and the melting index. The study demonstrated that these starches could effectively replicate melt-stretch properties in cheese analogs, improving the quality and consumer acceptance.

Another study by Fan et al. [17] formulated cheese analogs using 15% starch, 20% soy protein isolate, 7% gelatin, and 15% coconut oil, exploring various types of starch. Among the starches tested, the authors observed that wheat, tapioca, and waxy starches were the most promising, presenting similar effects on the product's structure, color, and flavor. Additionally, the plant-based cheese formulations exhibited a thicker texture and greater elasticity than traditional cheese, likely due to the starch forming complexes with fatty acids.

In addition to its use in cheese analogs, starch has also been utilized in yogurt-like beverages, as demonstrated by Pachekrepapol et al. [18]. Incorporating tapioca starch significantly enhanced the physicochemical and rheological properties of a coconut milk-based yogurt-like product. Its swelling properties effectively reduced syneresis, resulting in a creamier texture. Notably, adding tapioca starch did not adversely affect the viability of lactic acid bacteria during refrigerated storage.

Starch can also be used in meat analogs to improve water retention, simulate the fibrous texture of meat, and provide a mouthfeel that is like non-plant-based meat products. To simulate traditional meat cuts, fibrousness, and a meat-like structure are crucial for consumer acceptance. Processes like extrusion or shear cell processing are essential to creating plant-based products with this fibrous texture [2].

A study by Walsh et al. [21] found that blending whey protein with 20–40% starch by extrusion can create a fibrous texture and improve water retention in extruded protein. In contrast, another study by Zhang et al. [23] on plant-based meat analogs observed that the extrusion of wheat starch to peanut protein mixtures resulted in a softer, chewier texture that negatively affected the desired fibrous quality. The authors concluded that combining multiple polysaccharides with peanut protein during extrusion can enhance the fibrous structure of the final product.

Eggs are another product at the vanguard of research for plant-based alternatives. The development of egg analogs often encounters challenges, particularly in replicating key textural properties. Likewise, in cheese and meat analog products, a promising approach to improving the texture of egg analog products, like omelets, is incorporating starch, as a gelling agent. While studies by Lu et al. [25] demonstrated that the addition of agents like κ-carrageenan and gellan gum could not fully replicate the texture of traditional egg omelets, later research by Lu et al. [4] explored the potential of incorporating various starches, including corn, glutinous rice, and potato starch, to enhance the gelling properties of plant-based egg substitutes. Their findings revealed that a plant-based omelet with soy protein isolate and potato starch at ratio of 8:6 closely replicated the texture, particularly cohesiveness (plant-based: 0.75 and traditional: 0.79), and specific volume (plant-based: 1.28 and traditional: 1.32 $cm^3\ g^{-1}$) of a traditional egg omelet, offering broader applications in plant-based food systems, showing that starch are more suitable to achieve a texture like of traditional eggs.

In conclusion, the ample research into the use of starches in plant-based analog products emphasizes their significant role in simulating the texture and sensory attributes of traditional animal-based foods. Starches have proven effective as thickeners and stabilizers in dairy analogs, improving the creaminess of plant-based yogurts, cheeses, and ice creams, as well as enhancing melt-stretch properties in cheese alternatives. Similarly, in meat analogs, starches contribute to water retention and the fibrous texture crucial for consumer acceptance. The recent advancements in egg analogs further illustrate the versatility of starches, with studies showing that incorporating specific types, such as potato starch, can closely replicate the texture of traditional eggs. These findings underscore the importance of starches in advancing plant-based food technologies, offering promising solutions to create products that better meet consumer expectations and expand the range of plant-based options available in the market.

9.3 Functional Roles of Starch in Plant-Based Food Systems

9.3.1 The Influence of Starch and Its Interactions with Other Macromolecules on Plant-Based Food Properties

As previously described, starch is added to plant-based food systems for multiple purposes (increases yield and viscosity, modifies water holding capacity, gel formation properties, emulsification and surface hydrophobicity of protein molecules, adds body, provides fibrous structures and improves mouthfeel functionalities) [4, 26]. However, the role and impact of starch in such systems is highly dependent on different aspects, including its ratio amylose/amylopectin [27], the source [4], the particles' size [28], whether its native or modified (and, if modified, the process it has been subjected to) [2, 3], the presence of other macromolecules as ingredients (*e.g.* proteins, fat) [3], whether its endogenous or exogenous regarding the protein ingredient [2] and the technological processes and conditions it undergoes during the plant-based food manufacturing (*e.g.* temperature, time, shear) [3]. When designing plant-based systems, focusing on analogs, the aim is to resemble the animal-based product characteristics. Therefore, such aspects should be taken into consideration when defining which starch to use to achieve the desired texture, stability, and sensory qualities.

According to the literature review conducted, some authors examine the impact starch or mixtures of starch and macromolecules have when designing specific plant-based analogs, while others focus on the effect these ingredients have on numerous physicochemical and organoleptic properties, providing diverse data that could be helpful for the development of such products. Both approaches are described below.

Dobson and Marangoni [3] studied different starch-protein suspensions, using three pea protein isolates and three rapid swelling waxy maize starches; thermally inhibited (TI), chemically modified (CM), and granular cold-swelling (GCS). The viscosity of the suspensions at different protein:starch ratios indicate that, in general, when the concentration of starch is reduced, so does the viscosity, because this property is mainly influenced by the polysaccharide content. Nevertheless, for TI starch, an increase in viscosity is observed at high ratios of protein:starch, which indicates a synergistic effect due to electrostatic interactions and a small contribution from hydrogen bonds between the ingredients, because of TI conformation. This result is useful to design plant-based systems with high content of proteins, that have high nutritional and functional value, without compromising the texture [3]. Nilsson et al. [29], evaluated different ratios of starch:protein from faba beans and evaluated gel morphology/structure, observed that decreasing starch proportion resulted in lower viscosities during pasting, poorer water binding and hydration properties, and less firm gels and fibrous structure. This is caused by competition for water between proteins and starch; proteins take up more water and limit starch granule swelling, which in turn decreases the apparent viscosity. Moreover, proteins probably act as inert fillers, restraining the realignment of starch chains during gelation, and other authors have seen this effect is more pronounced for amylose-rich starches, because proteins are likely to interfere in the reorder of amylose [29, 30]. Although for plant-based systems, in general, a stable gel is desired, if a soft texture is preferred, less proportion of starch should be used.

While the latter studies provide an understanding of the possible behavior of starch, other ingredients, and processes commonly used in the formulation of analogs must be incorporated to better assess the real impact on the properties of these products. Moreover, some negative attributes observed when evaluating suspensions might be desirable when preparing analogs. For instance, for meat plant-based substitutes, Chen et al. [27] studied the effects of extrusion under high moisture (HME) conditions, of mixtures of pea protein isolate (PPI) and different ratios of amylose:amylopectin. During HME process, low viscosity is desirable to reduce energy consumption, and amylopectin made this possible. Moreover, when aiming to produce meat analogs, a fibrous, anisotropic structure is a key attribute for consumer's acceptance. In this regard, amylopectin enhanced the fibrous degree of the extrudates, while the increase of amylose gradually decreased this feature, measured as the ratio of the force required to cut a sample crosswise and lengthwise of the extrusion's direction [2]. Unlike the branched polymer, the linear chains of amylose may align to proteins during the HME, resulting in amylose-protein complexes that prevent the fibrous structure formation caused by protein polymerization. The results obtained indicate that a ratio of amylose:amylopectin of 0:1 is advantageous for PPI-based meat substitutes with a fibrous structure. This was also observed by Chen et al. [26], who characterized the structures of mixtures of PPI and amylose/amylopectin during each stage of HME processing. The authors concluded that, on one hand, amylopectin favored the unfolding of vicilin and legumin chains, the most important globular pea proteins [31], promoting protein rearrangement, aggregation and crosslinking that leads to a fibrous structure, while amylose that

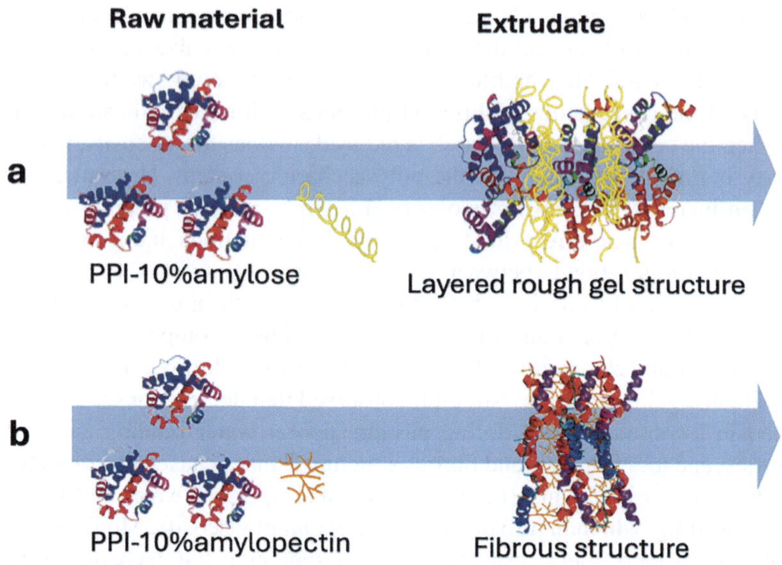

Fig. 9.2 Interactions between pea protein isolates and amylose (**a**) and amylopectin (**b**) molecules in an extruded product. (Adapted from Chen et al. [26])

filled the protein matrix enhanced the refolding of β-chains, which results in a layered rough gel structure (Fig. 9.2) [26]. Thus, to achieve a fibrous structure and improve the analog's tenderness, starches with high content of amylopectin (*e.g.* waxy corn starch) would be preferable.

Dobson et al. [3] created three doughs using TI, CM or GCS starches and PPI to mimic animal meat, through manual kneading, without the need for extrusion. Samples were analyzed with or without a final refrigeration step, which gave rise to cold-set and non-cold-set doughs, respectively, and texture results yielded significant differences among the three starches. In non-cold doughs, hardness, which must be high enough so that the dough is self-supporting even with the addition of other ingredients, increased with the addition of protein, which acts as a filler embedded in a starch matrix. CM was always the softest, probably because it is highly processed and has less granular structure. The starches contributed to the dough's resilience and cohesiveness, with lower values at higher protein concentrations, although for TI starches resilience was always higher. Chewiness was also improved with TI, and the authors concluded that the proteins act as active fillers, with strong interactions with the starch matrix. Overall, this study highlights the great differences obtained when varying the starch employed, with TI resulting in the best textural and mechanical properties for meat analogs.

Another plant-based substitutes that have gained attention in the last years are egg replacement ingredients. At the same time some companies focus on mimicking the taste and appearance of eggs (Oggs Aquafaba, Just egg, Beyond egg, etc.) [25] considerable research is being conducted to obtain products with similar textural and emulsification properties, and well as cooking functionalities as eggs [32], for further development of egg-free mayonnaises, dressings, cakes, noodles, omelets, etc. Lu et al. [25] studied different formulations for omelet production, using chickpea flour as source of starch, soy protein isolate, different quantities of κ-carrageenan (κ-C) and high acyl gellan gum (GG), and other ingredients, to simulate egg properties. The resulting product using 0.3% κ-C best matched some properties of eggs (hardness, specific volume, and gel strength), with starch contributing to gelling performance, due to starch gelatinization and κ-C's entanglement with starch and proteins. κ-C-protein electrostatic interactions in 0.3% κ-C led to the formation of a composite gel matrix with starch filled in although additional functionalities of liquid egg require further investigation. Lu et al. [4] studied the effect of potato starch, corn starch, and glutinous rice starch, at different soy protein isolate (SPI)/starch ratios, on physicochemical properties of plant-based omelets and gels. The results indicated that variations in interactions SPI-starch could be attributed to the origin of the starches, *e.g.* Potato starch is richer in phosphorous that promotes starch chains' separation and consequently more water access during gelatinization, as well as more protein-starch electrostatic interactions. Moreover, the abundance of hydroxyl groups in potato starch resulted in greater protein-starch bonding that yielded a stronger structure, and high water-binding capacity, contributing to cohesiveness and springiness of the system. In general, increasing starch proportion increased hardness, springiness, and chewiness of omelets, while cohesiveness followed differences trends, with SPI/potato starch 8:6 that matched the texture (cohesiveness and specific volume) of egg omelets. Nevertheless, more research should be conducted to improve the mouthfeel properties of the resulting product.

Dairy-free cheese analogs are no exception to the extensive research on various plant-based alternative foods. Dobson and Marangoni [33] compared a high-protein plant based cheese using waxy starch, with commercial dairy and plant-based cheese alternatives. Native waxy starch improved the viscoelastic properties of the matrix, as proteins and fat modulated gelatinization. These macromolecules acted as particulate fillers, increasing the network strength and promoting strategic connection points during starch retrogradation. As a result, it was possible to obtain an analog with similar hardness, meltability and spreadability to commercial dairy cheeses, with meltability related to the proportion of amylopectin (higher in waxy starches), the presence of proteins and the production process. The main disadvantage of the formulation was that high amounts of oil loss occurred during melting. This highlights the importance of the type of starch and presence of other macromolecules to reach the desired properties. The oil loss problem was addressed in a subsequent study by the same authors [33], through the incorporation of low concentrations of oil gelators. The ability to stretch when melted (*i.e.* melt-stretch properties) was also studied by Yun et al. [6], combining machine learning with rheological results. When analyzing nine different starches pastes, it was concluded that changes in structural characteristics of the starch samples (amorphous and crystalline forms) before and after pasting, were linked to their melt-stretch lengths, and this was confirmed when incorporating waxy potato starch to a plant-based analog [6].

To sum up, the previous studies highlight the impact different starches have on a wide variety of textural, rheological, and sensory attributes when designing plant-based analogs. This means that there is not one type of starch suitable for all products, and that the decision on which one to include, will depend on the other ingredients used, and their proportions, the technological processes and the desired properties of the final product.

9.3.2 Impact on Nutritional Profile

Shifting toward plant-based diets is believed to offer health, environmental and ethical advantages [29, 34]. Nevertheless, the health benefits of plant-based products are highly dependent on the quality of the ingredients and the technological process. Moreover, when designing analogs, some producers only focus on mimicking the texture, consistency, and other physicochemical properties of the animal-based product, and not on the nutritional value [9]. When it comes to starch addition, numerous studies describe potential health benefits of this polymer. In general, these benefits are related to the incorporation of resistant starch and the use of starch as fat replacer. For instance, Rashwan et al. [28] describes a technological process that yields a starch with slower speed of digestion compared to a standard, due to the high content of amylose. Amylose is more organized and densely packed than amylopectin, which results in the formation of a physical barrier that hinders the digestive enzyme action and therefore reduces its digestibility. Consequently, glucose is released more slowly, and the glycemic index of the product could be reduced. The effect on slow digestion rate is also described by Chen et al. [35] that remarks that proteins, starch (resistant or not) and lipids, which are the main ingredients in meat analogs, form complexes that augment resistance to amylolysis. Rashwan et al. [28] also reports that resistant starch positively modulates the gut microbiome through the stimulation of *Bifidobacterium* and *Akkermansia*, both genera that play important roles in gut health and overall metabolism. Cione et al. [36] highlights that resistant starch not only has low glycemic index (compared to non-resistant starch), but also indirectly reduces blood lipid levels. When it comes to the application of starch as lipid replacer, Chen et al. [37, 38] and Huang et al. [38] indicate that carbohydrates, such as native or modified starches, are useful to mimic fat globules due to their physicochemical, textural and organoleptic properties, while providing less energy than lipids.

Regarding the nutritional drawbacks of starch addition, Lu et al. [24] optimized a plant-based egg analog and observed that the bioaccessibility of amino acids of the formulae of starch that best emulated the egg omelet properties, was reduced, affecting negatively protein digestibility. This impacts negatively not only on the amount of peptides and amino acids available in the bloodstream for maintenance of health, but also promotes the formation of biogenic amines in the gut by the microbiome, which promotes bowel diseases [39]. Still, there is limited information regarding the effect of starch on proteins' (or other nutrients) bioaccessibility, because in general studies do not evaluate the impact on such molecules due to specific starch interactions, and the negative effects might be related to other factors, such as the origin of the nutrient, *e.g.* plant-based proteins that are less [40], the production process and/or interactions with other ingredients. Another negative aspect of starch addition, caused by processing, is the possible increased Maillard reaction when proteins are also present, and the presence of flatulence-causing oligosaccharides and anti-nutritional factors, when using specific starch sources (*e.g.* faba bean starch) [2].

Overall, even though authors report different effects of starch addition on the nutritional profile of plant-based analogs, either positive or negative, the potential impact on health should be determined for each product individually, because interactions with other macromolecules, the origin of ingredients and starch itself, if it is resistant or not, the processing method (*e.g.* high temperature, extrusion, fermentation), among other factors have a significant influence on the health effect of the final product.

9.4 Starch in Plant-Based Foods During Processing, Cooking and Storage

As previously reported, starch is vital in determining plant-based foods' texture, stability, and quality. Its behavior under various processing and storage conditions significantly influences the sensory and nutritional properties of the final product. Understanding how starch contributes to the stability of those products is crucial for optimizing food processing methods and improving the shelf life and quality of plant-based products. This section provides an in-depth analysis of starch stability across different contexts, including processing, storage, and cooking.

9.4.1 Stability of Plant-based Products During Cooking and Processing

The behavior of starch, especially its processes of gelatinization and retrogradation, is crucial in shaping the physical properties of plant-based alternatives. A thorough understanding of how those processes can optimize meltability, gel strength, and sensory attributes during production and cooking is essential to modulate new plant-based formulations. This section explores how starch can improve the functional properties and stability of the most important plant-based foods groups (dairy, egg and meat analogs) during cooking and processing, emphasizing its influence on the final texture, quality, and overall functionality of the products.

In plant-based cheese systems, achieving the desired texture and meltability during cooking is critical for consumer acceptance. Meltability is a function of how heat interacts with starch and other components during cooking, influencing the transition of the product from a solid to a more viscous state [41]. Understanding the interactions of various ingredients during cooking, especially the behavior of starch, is essential for optimizing meltability and texture. A study by Lyu et al. [41] investigated the effects of emulsion and protein incorporation on the structural and melting properties of starch gels designed as model systems for plant-based cheese. The findings revealed that while native maize starch gels did not melt, oxidized potato starch gels melted completely above 70 °C, exhibiting lower viscosity than molten dairy cheese. Incorporating a protein-stabilized sunflower oil emulsion created an oil droplet network that partially inhibited melting. Heated pea protein also formed a network that limited melting, resulting in an inhomogeneous appearance due to protein aggregation. At lower protein levels, syneresis occurred, but adding protein without heat treatment prevented melting and maintained a uniform appearance. Combining emulsion and protein at higher levels produced more homogeneous gels and reduced melting. This study highlights the potential of starch as a key component in developing dairy cheese substitutes. It also emphasizes the importance of starch stability during processing to achieve desirable melting properties.

Another study [9] emphasized the role of starch in the meltability of cheese analogs. The authors attribute the unique meltability of high-protein plant-based cheese alternatives to both formulation and mixing methods. This high-protein cheese utilizes native waxy corn starch as the sole starch component, with protein incorporated before heating. This initial swelling of the protein is followed by the starch. As the mixture is gradually heated and mixed, the starch swells and gelatinizes, creating a viscous system that effectively accommodates the protein. The continuous starch phase establishes a stable network that keeps the fat dispersed. The mixing process includes short segments of high shear, which visibly reduce viscosity and increase stickiness, likely causing granule breakdown and fragmentation. Crucially, the mixing temperature is maintained below 80 °C to prevent protein denaturation and excessive network formation. After mixing, the viscous system is left to set in refrigeration, allowing it to solidify into a gel. The resulting high-protein plant-based cheese alternative features a unique structure with dispersed protein and fat, significantly enhancing its melt properties.

Similarly, starch is key in plant-based egg systems to achieve desirable gelling performance and texture during frying [4, 25]. Gelatinization is essential for forming a gel network that mimics the properties of liquid eggs. When heated, starch absorbs water, swells, and disrupts its crystalline structure, transforming into a viscous gel that contributes to the overall consistency of the product [4, 25]. A study by Lu et al. [25] observed that the incorporation of hydrocolloids like κ-carrageenan (κ-C) and gellan gum (GG) enhances the gelling process of egg analogs produced with chickpea flour and soy protein isolate. This enhancement occurs due to the entrapment of hydrocolloids with starch (from flour) and proteins, resulting in a more robust gel network. It exhibits similar gelling performance to the commercial liquid egg (plant-based), mainly due to the gelatinization of starch and the formation of proteins in the gel network. The fact observed by the authors is interesting since the sample's stronger gel strength and less viscosity is associated with its higher air-holding capacity during frying, mimicking animal-based eggs better. Overall, the pronounced gelatinization of starch in these plant-based egg systems is crucial for creating strong, stable gels. This demonstrates the significant benefits by improving texture and functional properties in plant-based food during processing and frying.

Water loss during the cooking of meat analogs is a critical issue that impacts the texture and quality of beef and plant-based patties. A study by Vu et al. [42] examined the effects of different cooking methods (pan frying, air frying, and oven baking) on the physicochemical properties of commercial beef and plant-based burgers, specifically focusing on texture, appearance, cook loss, and shrinkage. The authors observed that cooking loss increased with cooking time, with beef patties experiencing significantly higher fluid loss than plant-based meat analogs The authors concluded that plant-based patties retained more water due to a stable protein network and the inclusion of modified starch and methylcellulose, which formed a porous matrix that minimized cook loss and maintained texture. The textural changes of both plant-based and non-plant-based meat were also assessed. In plant-based burgers, modified starch plays a crucial role by swelling and forming a 3D network that traps moisture, helping to retain texture during cooking. Overall, while significant differences were observed between beef and plant-based patties, the cooking method had a minimal impact on their final properties. The structural differences, particularly modified starch in plant-based patties, were key factors influencing cook loss and texture.

In conclusion, starch is fundamental to improve stability, texture, and overall quality of plant-based food products during cooking and processing. Its ability to undergo gelatinization and form a gel network is critical in mimicking the functional properties of animal-based foods, such as meltability in plant-based cheese and gelling performance in egg analogs. In addition, starch interaction with proteins and hydrocolloids further enhances the functional properties of meat analogs, contributing to improved water retention, structure, and texture. Moreover, the stability of starch during different processing techniques, whether in frying or cooking, is crucial in maintaining the product's integrity, ensuring a desirable texture, and preventing issues like syneresis.

9.4.2 Stability of Plant-based Products During Storage

The stability of plant-based products is essential for ensuring quality, shelf life, and consumer satisfaction. As demand for plant-based alternatives grows, the food industry faces challenges in preserving texture, flavor, and structural integrity during storage. Unlike animal-based ingredients, plant-based formulations rely on alternative proteins, fats, and starches that are more sensitive to temperature, humidity, and time factors. Issues like phase separation, moisture migration, and oxidation can compromise the stability of emulsions, gels, and solids. To ensure consistent product performance, exploring ingredient modifications, advanced processing techniques, and optimized packaging is crucial.

In developing meat replacers, a study by Osman et al. [43] examined the effect of acetylated corn starch on the physicochemical, textural, sensory, and microstructural characteristics of plant-based beef patties during frozen storage (−20 °C) for 60 days. The use of acetylated corn starch (15%) significantly improved the stability of patties, reducing cooking loss, diameter reduction, and dimensional shrinkage, indicating better moisture retention and structural integrity. Additionally, it maintained thickness and prevented excessive firmness, enhancing texture. The pH remained stable, improving tenderness, color, and sensory attributes, further contributing to long-term stability.

As the demand for plant-based alternatives continues to grow, the food industry faces the challenge of developing emulsifiers that can effectively replace traditional egg-based ingredients in products like mayonnaise. In a study by Taghavi et al. [5], pre-gelatinized rice starch enhanced the emulsifying capacity and stability of a pickering emulsion system. At a low concentration (200 mg/mL of oil), pre-gelatinized rice starch achieved a 100% emulsion index and the smallest droplet size (17 μm). Its long-term stability over 14 days outperformed both native and OSA-modified starches. The authors proposed that gelatinization disrupts the crystalline structure of starch, increasing water absorption and swelling, which improves emulsifying efficiency and stability over time. These findings highlight the potential of pre-gelatinized rice starch as a highly effective plant-based emulsifier, providing a viable alternative to traditional egg-based emulsifiers. Its use not only enhances plant-based mayonnaise but also improves storage stability in products like salad dressings, sauces, non-dairy creams for baking, and other plant-based items stored at room temperature.

Ensuring the stability of cheese analogs over time and temperature variations is crucial for maintaining quality and consumer acceptance. In this context, a study by Dharaiya et al. [44] evaluated mozzarella cheese analogs made with pre-gelatinized starch and either acid casein, rennet casein, or a combination, against non-plant-based mozzarella (control) during refrigerated storage at 7 °C. Over 28 days, the analogs demonstrated superior shred ability, meltability, stretch, and reduced fat leakage compared to the control. The gelatinized starch likely enhanced stability by improving moisture retention, minimizing phase separation, and preserving texture during storage. Given its enhanced storage stability and baking properties, the authors recommend this plant-based ingredient for refrigerated ready-to-use pizza toppings.

Given their high-water content, plant-based meat analogs are typically preserved by freezing, as low temperatures inhibit microbial growth. However, this process can result in ice crystal formation, which may disrupt the fiber network structure and compromise the quality and sensory properties of the final product [45]. This outcome was well documented by Jimenez-Colmenero et al. [46], which examined the microstructural changes in konjac gels with pre-gelled cornstarch as fat replacers in reduced-fat meat products during chilled storage and freezing/thawing cycles. The freezing/thawing process caused noticeable changes, including a sponge-like texture and a 10% decrease in water binding capacity, with syneresis measured at 4%. Water binding capacity remained unchanged after 20 days of frozen storage, and syneresis was stable. These results indicate that freezing/thawing significantly reduces the stability of konjac gels, affecting water retention and the polymer-solvent relationship. Although pre-gelled cornstarch likely improved texture and moisture retention, ice crystal formation reduced water retention within the gel network, causing dislocation not fully reversed upon thawing. Despite advancements in the development of meat analogs, the existing literature lacks detailed discussions on the storage stability of these products, highlighting a significant gap in research. Exploring how different storage conditions impact the quality attributes of plant-based meat analogs presents a valuable opportunity for further investigation.

In conclusion, the ongoing exploration of stability in plant-based products is vital for meeting consumer expectations and ensuring quality over time. As the market for plant-based alternatives expands, further research into the storage stability of these products will provide valuable insights, paving the way for improved formulations that retain their quality and appeal over time.

9.5 Sensory Properties of Starch Plant-based Products

Plant-based foods desire to meet consumer expectations by replicating the sensory characteristics of their animal-based counterparts. Yet, they often fall short regarding mouthfeel, texture, and overall satisfaction. As the demand for these alternatives grows, so does the need for innovative ingredients that can improve these sensory attributes while contributing to the product's nutritional profile.

Starch is a versatile and functional ingredient that offers a promising solution for developing new plant-based products. Its ability to modify texture, improve stability, and mimic the mouthfeel of fats or proteins makes it a valuable tool for meeting sensory expectations and enhancing nutritional value [2, 3, 6, 19]. By incorporating starch, manufacturers can create plant-based products that appeal to a broader range of consumers without compromising taste, appearance, or health benefits [2, 4, 6].

Research has shown that both native and modified starches can effectively achieve the desirable sensory qualities of traditional dairy products and, at the same time, improve the functional properties of the final product. Techniques for modifying starches have revealed improved properties, such as increased water solubility, excellent film-forming ability, lower viscosity, and reduced retrogradation tendency. These characteristics make starches highly functional and adaptable ingredients in food formulations [47, 48].

For instance, a study by Ahsan et al. [47] investigated using oxidized potato starch as a substitute for fat and casein in imitation mozzarella cheese. The results demonstrated that oxidized (a modified starch type) potato starch successfully replaced both components without compromising the cheese's sensory profile. These properties contribute to the cheese's desirable sensory characteristics and ensure that consumers would likely not notice any perceptible changes, making the cheese comparable to the original product.

Another study by Zehra et al. [48] assessed the sensory attributes of analog cheese and found that samples containing native and hydroxypropylated starch extrudates exhibited superior stretchability and melting than those in the control sample (cheese without starch), aligning with rheological measurements. Additionally, the cheese with hydroxypropylated starch extrudates did not show significant differences in appearance, aroma, or texture compared to the control.

A study by Lorusso et al. [20] highlighted how starch influenced the sensory profiles of yogurt-like beverages fermented with different probiotic types (*Lactobacillus plantarum, Weissella confusa,* and *Lactobacillus rhamnosus*). Among the probiotics tested by the authors, *Lactobacillus plantarum* produced the most intense odor and flavor, while *Weissella confusa* resulted in a sweeter taste and lower perception of odor and astringency. In addition, the authors observed that, the production of exopolysaccharides by *Lactobacillus plantarum* and *Lactobacillus rhamnosus* contributed to yogurt-like sour and dairy perceptions, with increased particle perception but no significant differences in appearance. Starch, particularly from quinoa, was crucial in shaping these sensory qualities. Acting as a substrate during fermentation, it facilitated the production of flavor compounds and helped create a creamy texture like that of fermented dairy products. Additionally, quinoa starch contributed to the overall textural attribute of the beverages, enhancing their sensory appeal without compromising the quality.

In another study, a gluten- and lactose-free plant-based yogurt-like was produced as the base for a novel ice cream, where starch played a crucial role in enhancing sensory and structural qualities. Starch gelatinization during thermal treatment created a creamy, dense texture like dairy-based yogurt. This modification improved smoothness, stabilized consistency, and provided structural integrity, allowing the ice cream to maintain its form over time. These features helped the plant-based ice cream closely replicate the sensory experience of dairy products, with only minor differences in taste, appearance, and texture [10].

In summary, these results underscore the potential of native and modified starches to maintain or enhance the sensory qualities of dairy analogs, confirming that they can be effectively used to create plant-based products that meet consumer expectations without negatively impacting taste, texture, or overall acceptability.

Plant-based meat alternatives enclose various products designed to replicate the characteristics of traditional meat, including meat substitutes, analogs, and imitation meats made entirely from plant proteins. These products aim to deliver a sensory experience like that of animal meat, offering features such as fibrous texture, moisture levels, and chewing resistance. Despite their plant-based origin, creating plant-derived fibers that closely mimic the texture and appearance of meat remains a significant challenge, driven by the need to meet the taste and texture expectations of meat consumers [2, 3, 49]. Meat analogs incorporate plant polysaccharides such as sugars, dietary fiber, and starch to enhance protein-water-lipid interactions during extrusion, ultimately contributing to the development of a meat-like fibrous texture with enhanced aesthetic and structural characteristics [2], as previously mentioned in Sect. 9.4.

The impact of vegetable proteins extruded with starch on the sensory qualities of meat analogs was investigated by Walsh et al. [21]. Sensory evaluation of those analogs indicated that beef patties analogs extended with whey protein were preferred over those extended with vegetable protein. As previously reported, Sect. 9.4, the patties with whey protein had better cooking loss, fat retention, and less diameter shrinkage compared to those with textured vegetable protein, reflecting the improved structural and sensory qualities contributed by whey protein and starch. Extruded meat analogs produced using whey protein concentrate and cornstarch demonstrated an acceptable appearance and texture, with consumer scores averaging above 6 ("like slightly"), suggesting that the starch played a key role in enhancing the product's sensory appeal and structural integrity, making it a viable option for plant-based meat alternatives.

Starch can also improve the functional attributes of egg analogs without compromising the sensory profile of this product. A recent study by Lu et al. [24] assessed the sensory attributes of egg omelets produced with a modified starch—hydroxypropyl methylcellulose (HPMC), soy protein isolate, potato starch, sunflower oil, chickpea flour, baking powder, transglutaminase, κ-carrageenan, and potassium chloride. The authors observed that adding 1.5% HPMC reduced the pastiness of the vegan omelet. This improvement was attributed to the significant starch content (6%) in the formulations, which, combined with HPMC, played a crucial role in the thermal gelation process. The gelatinization of starch during cooking contributed to a smoother, creamier texture by enhancing moisture retention and oil adsorption. The 1.5% HPMC formulation closely mimicked the sensory qualities of an egg omelet in terms of fluffiness, softness, smoothness, crispiness, and silkiness, which positively impacted the overall mouthfeel compared to the formulation without HPMC.

To sum up, starch is essential for enhancing plant-based foods' sensory and functional qualities. Its ability to improve texture, moisture retention, and overall stability makes it a key ingredient in developing products that closely mimic their animal-based counterparts. By harnessing starch's unique properties, manufacturers can create more appealing and acceptable plant-based alternatives, thus meeting consumer expectations and fostering broader market acceptance. Ongoing research into starch applications will further unlock its potential, paving the way for innovative solutions in the plant-based food sector.

9.6 Challenges and Future Prospects for Starch Utilization

9.6.1 Technological and Regulatory Challenges

Many studies have been carried out in recent years to understand the fundamental structural characteristics of starch that produce certain functions. However, there are still no complete explanations as to why small changes in the structure alter functionalities, as well as what is responsible for governing certain properties of starch

[50]. In addition, these changes vary according to the application of starch in plant-based foods, such as plant-based meat, milk alternatives, food packaging, and as a stabilizer in foods [28].

In general, understanding how the linear chains and branches of amylopectin are organized among the two models of internal structure (tree-shaped cluster and backbone) would be better according to the applications for specific uses from the physical point of view (enzymatic and chemical modifications of starch), as well as genetics and improvement [50].

Due to great global interest, products mimicking meat or offering protein-rich alternatives have grown. However, they also present some challenges in their production, mainly regarding texture, flavor, and color [2, 8]. According to McClemnts et al. [5], in addition to flavor and texture issues, there are difficulties in manufacturing products with appearance, fluid retention, cooking, and mouthfeel similar to animal proteins.

Starch is generally added to meat substitute products as a functional ingredient or filler to improve textural properties or to retain water [2]. In general, adding starch tends to increase the toughness of plant-based products (meat), and by adequately increasing the ratio of straight-chain to branched-chain starch, the fiber structure can be improved [51]. However, when vegetable oils replace animal fats to improve the lipid profile in plant-based meats, starch plays an important role, helping to maintain the product's qualities and sensory aspects [28]. In general, the main benefits of starch, in addition to its cost-effectiveness, are to adjust functional traits, mainly cold swelling capacity, water solubility, and rheological properties through modifications [52].

In the case of fully plant-based milk, starch (and other biopolymers) is generally added as thickening agent to create desirable textural attributes in the mouth or to inhibit the sedimentation of insoluble matter and the creaminess of lipid droplets [53]. For plant-based drinks made from starchy materials (such as cereals or pseudo-cereals), enzymatic hydrolysis is generally necessary since, at high temperatures, the starch begins to gelatinize, making the drink more viscous [54, 55].

Regarding cheese analogs, starch is the most common ingredient used as a polysaccharide base to form the gelled structures. However, these products contain compounds that can be harmful to human health if consumed regularly, such as high amounts of saturated fatty acids and refined carbohydrates. In addition, they do not provide the same beneficial nutrients as regular cheese, such as calcium, protein, vitamin A, and vitamin B12 [53].

The main way to better utilize starch in plant-based foods will likely require adapting the starch depending on the added product. Starch modifications have been extensively studied, especially physical and enzymatic modifications [56]. In the case of enzymatic modifications, these allow modifications in the starch structure without collateral damage, improving starch's solubility and thermal properties and increasing its applicability. In the case of physical modifications, they improve water retention and resistance to retrogradation, in addition to improving starch functionality and digestibility by 20% [28].

There are still many discussions about the regulation of plant-based food products, which vary according to the region and should be considered with a view to their development and commercialization [52]. However, where there is still no regulation of these plant-based products and beverages, it is necessary to understand the nutritional profiles of animal origin that will be replaced [57]. Since regulating plant-based products is still being discussed and implemented, adding starch has not been addressed. Therefore, this compound must be added scientifically to improve the desired characteristics of the product without harming consumers.

9.6.2 Future Innovations and Research Directions

As previously mentioned, starch modification has increased its applicability to different plant-based foods. In addition to the conventional (heat-moisture treatment) and enzymatic methods, some emerging physical methods of starch modification have been studied, such as high hydrostatic pressure (HHP), cold plasma, microwave, pulsed electric field, and gamma-irradiation [58]. Maniglia et al. [59] also complemented this with ozone, ultrasound, and high homogenization processing technologies.

Many plant-based foods have been developed containing modified starches in their composition. Among them are mayonnaise with modified rice starch granules as a vegetable emulsifier [5], a plant-based meat analog with a starch matrix filled with protein particles and fibers [3] fermented soy beverage to improve quality by controlling solubility and viscosity, and preventing syneresis [60], and others (yogurt, cheese, and fish fillets) [61–63].

Although physical modifications of starch have shown great potential, there is still much to be elucidated regarding these techniques. For modification with HHP, using non-plating solvents (such as oils and ethanol) may be an alternative to help understand the changes in starch caused by pressure. In the case of cold plasma, using different facilities leads to varying results, and chemical complexity and insufficient penetration are the main challenges to overcome. Regarding microwaves, eliminating thermal effects has been controversial for decades [58].

In the case of ozone, it is still necessary to apply these compounds formed after ionization with other food ingredients (*e.g.*, proteins, fats) to evaluate their potential and scalability. As for the ultrasound technique, the parameters must be standardized to avoid equipment erosion with increased temperature, which can negatively affect processing [59].

Due to extensive research in recent years, we are closer to resolving some basic questions about the starch structure and its influence on the functionalities of new products, making them more consistent and innovative [50]. However, studies are still needed to adapt the structure and properties to achieve such functionality, that is, a detailed understanding of which property is desired (in which range) and which process (and parameters) is necessary to achieve it [59].

9.7 Conclusions

Starch is a vital ingredient in plant-based food systems, showcasing significant potential to replicate animal-based foods' functional and sensory characteristics. Recent research highlights its versatility in meat, dairy, and egg analogs, where it enhances yield, viscosity, water-holding capacity, gel formation, emulsification, and protein hydrophobicity. In dairy alternatives, starch contributes to creaminess and melt-stretch properties, while in meat analogs, it improves water retention and fibrous texture, both crucial for consumer acceptance. Additionally, certain starches, such as potato starch, effectively mimic egg textures, underscoring their importance in developing plant-based products that meet consumer expectations for texture and flavor.

The choice of starch type must be tailored to specific products, as ingredient proportions, processing methods, and desired characteristics significantly affect starch functionality. Additionally, starch interactions with other macromolecules are crucial in determining plant-based systems' texture, stability, and quality. Its ability to undergo gelatinization and form a gel network is essential for mimicking the functional properties of animal-based foods, such as meltability in plant-based cheese and gelling performance in egg alternatives. Maintaining the stability of starch-based plant systems during various processing techniques, including frying and cooking, is critical for preserving product integrity, achieving desirable textures, and preventing issues like syneresis. Furthermore, the nutritional impact of starch can vary based on its origin, resistance levels, processing conditions, and interactions with other ingredients, highlighting the need for product-specific studies to assess its health implications accurately.

In summary, starch plays a critical role in enhancing plant-based products' stability, texture, and sensory attributes, facilitating the development of alternatives that closely mimic animal-derived foods. Ongoing research into starch modification techniques, such as high hydrostatic pressure, cold plasma, and ultrasound, is vital for expanding its applicability in plant-based food systems. Continued advancements in starch technology will foster the creation of innovative, high-quality, plant-based options that meet market demands and comply with regulatory standards.

Further investigation into the nutritional properties of starch in plant-based food systems is essential. While starch is important for these products' texture and sensory qualities, its implications for human health are still not fully comprehended. Additional studies are needed to evaluate how different types of starch, and their modifications affect digestibility, nutrient absorption, and overall consumer health. Gaining this knowledge will facilitate the formulation of plant-based products that satisfy sensory expectations and provide nutritional benefits, aligning with the increasing market demand for healthy and sustainable options.

Acknowledgments G.D. Sorita acknowledges the support of the National Council for Scientific and Technological Development (CNPq, 150281/2024-2). W.D.C. Chacon gratefully acknowledges the Fundação de Amparo à Pesquisa e Inovação do Estado de Santa Catarina (FAPESC) for the PhD fellowship (3003/201). M. Rajchman wants to thank the National Agency for Research and Innovation (Uruguay) (POS_EXT_2023_1_17530). E. Leonarski acknowledges the support of the National Council for Scientific and Technological Development (CNPq process: 150786/2024-7.).

Conflicts of interest The authors declare no conflict of interest.

References

1. Kraak VI, Aschemann-Witzel J. The future of plant-based diets: aligning healthy marketplace choices with equitable, resilient, and sustainable food systems. Annu Rev Public Health. 2024;45(1):253–75.
2. Bühler JM, Schlangen M, Möller AC, Bruins ME, van der Goot AJ. Starch in plant-based meat replacers: a new approach to using endogenous starch from cereals and legumes. Starch - Stärke. 2022;74(1–2):2100157.
3. Dobson S, Laredo T, Marangoni AG. Particle filled protein-starch composites as the basis for plant-based meat analogues. Curr Res Food Sci. 2022;5:892–903.
4. Lu Z, Liu Y, Lee YEJ, Chan A, Lee PR, Yang H. Effect of starch addition on the physicochemical properties, molecular interactions, structures, and in vitro digestibility of the plant-based egg analogues. Food Chem. 2023;403:134390.
5. Taghavi E, Andriani C, Nordin N, Awang Seruji AZR, Wan Rasdi N, Abdul HN. Rheological and stability of mayonnaise-based pickering emulsions stabilised by modified rice starch granules as a plant-based emulsifier. Int J Food Sci Technol. 2024;59(8):5651–63.
6. Yun S, Jeong S, Lee S. Machine learning-combined rheological and structural characterization of starches for enhanced melt-stretch features in plant-based cheese analogues. Food Hydrocoll. 2024;157:110456.
7. Deziderio MA, de Souza HF, Kamimura ES, Petrus RR. Plant-based fermented beverages: development and characterization. Food Secur. 2023;12(22):4128.
8. Alcorta A, Porta A, Tárrega A, Alvarez MD, Pilar VM. Foods for plant-based diets: challenges and innovations. Foods. MDPI AG. 2021;10:293.
9. Dobson S, Marangoni AG. Methodology and development of a high-protein plant-based cheese alternative. Curr Res Food Sci. 2023;7:100632.
10. Pontonio E, Montemurro M, Dingeo C, Rotolo M, Centrone D, Carofiglio VE, et al. Design and characterization of a plant-based ice cream obtained from a cereal/legume yogurt-like. LWT. 2022;161:113327.
11. Dorantes-Fuertes MG, López-Méndez MC, Martínez-Castellanos G, Meléndez-Armenta RÁ, Jiménez-Martínez HE. Starch extraction methods in tubers and roots: a systematic review. Agronomy. 2024;14(4):865.
12. Zheng X, Wang Q, Li L, Liu C, Ma X. Recent advances in germinated cereal and pseudo-cereal starch: properties and challenges in its modulation on quality of starchy foods. Food Chem. 2024;458:140221.
13. Fang F, Tuncil YE, Luo X, Tong X, Hamaker BR, Campanella OH. Shear-thickening behavior of gelatinized waxy starch dispersions promoted by the starch molecular characteristics. Int J Biol Macromol. 2019;121:120–6.
14. Seung D. Amylose in starch: towards an understanding of biosynthesis, structure and function. New Phytol. 2020;228(5):1490–504.

15. Wang Y, Ou X, Al-Maqtari QA, He HJ, Othman N. Evaluation of amylose content: structural and functional properties, analytical techniques, and future prospects. Food Chem X. 2024;24:101830.
16. Mohamed IO. Effects of processing and additives on starch physicochemical and digestibility properties. Carbohydr Polym Technol Appl. 2021;2:100039.
17. Fan M, Wei T, Lu X, Liu M, Huang Y, Chen F, et al. Comprehensive quality evaluation of plant-based cheese analogues. J Sci Food Agric. 2023;103(13):6595–604.
18. Pachekrepapol U, Kokhuenkhan Y, Ongsawat J. Formulation of yogurt-like product from coconut milk and evaluation of physicochemical, rheological, and sensory properties. Int J Gastron Food Sci. 2021;25:100393.
19. Kiss KFDC, Oliveira ÉR, Ribeiro AEC, Oliveira AR, de Vinhal GLRRB, de Oliveira TF, et al. Texture profile of fermented rice extracts with probiotic strains and different contents of waxy maize starch, and sensory acceptance of flavoured selected extract. Int J Food Sci Technol. 2020;55(2):490–9.
20. Lorusso A, Coda R, Montemurro M, Rizzello C. Use of selected lactic acid bacteria and quinoa flour for manufacturing novel yogurt-like beverages. Food Secur. 2018;7(4):51.
21. Walsh MK, Nam SH, Pettee BC, Carpenter CE. Characterization of textured whey protein produced at varying protein concentrations. J Food Process Preserv. 2008;32(3):503–16.
22. Wu Y, Li Z, Yang L, Zheng H, Xue C. Construction of 3D printed salmon fillet simulants: improving printing performance by blending corn starch and flaxseed oil with pea protein and post-printing texturization via transglutaminase. Food Hydrocoll. 2024;155:110242.
23. Zhang J, Liu L, Jiang Y, Shah F, Xu Y, Wang Q. High-moisture extrusion of peanut protein–/carrageenan/sodium alginate/wheat starch mixtures: effect of different exogenous polysaccharides on the process forming a fibrous structure. Food Hydrocoll. 2020;99:105311.
24. Lu Z, Lee PR, Yang H. Using HPMC to improve sensory properties of vegan omelet analogue: effect of HPMC on water retention, oil adsorption, and thermal gelation. Food Hydrocoll. 2023;144:108938.
25. Lu Z, Lee PR, Yang H. Chickpea flour and soy protein isolate interacted with κ-carrageenan via electrostatic interactions to form egg omelets analogue. Food Hydrocoll. 2022;130:107691.
26. Chen Q, Zhang J, Zhang Y, Kaplan DL, Wang Q. Protein-amylose/amylopectin molecular interactions during high-moisture extruded texturization toward plant-based meat substitutes applications. Food Hydrocoll. 2022;127:107559.
27. Chen Q, Zhang J, Zhang Y, Meng S, Wang Q. Rheological properties of pea protein isolate-amylose/amylopectin mixtures and the application in the high-moisture extruded meat substitutes. Food Hydrocoll. 2021;117:106732.
28. Rashwan AK, Younis HA, Abdelshafy AM, Osman AI, Eletmany MR, Hafouda MA, et al. Plant starch extraction, modification, and green applications: a review. Environ Chem Lett. 2024;22(5):2483–530.
29. Nilsson K, Johansson M, Sandström C, Eriksson Röhnisch H, Hedenqvist MS, Langton M. Pasting and gelation of faba bean starch-protein mixtures. Food Hydrocoll. 2023;138:108494.
30. Bravo-Núñez Á, Garzón R, Rosell CM, Gómez M. Evaluation of starch–protein interactions as a function of pH. Foods. 2019;8(5):155.
31. O'Kane FE. Molecular characterisation and heat-induced gelation of pea vicilin and legumin. Wageningen: Wageningen University; 2024.
32. Ortiz-Solà J, Chic A, Martín-Gómez H, Aguiló-Aguayo I. Scientific insights into the vegan egg: composition, characteristics, and practical applications. In: Handbook of plant-based food and drinks design. Elsevier; 2024. p. 183–97.
33. Dobson S, Marangoni AG. Fat stabilization techniques for the reduction of oil loss in high protein plant-based cheese. Food Hydrocoll. 2024;156:110362.
34. McClements IF, McClements DJ. Designing healthier plant-based foods: fortification, digestion, and bioavailability. Food Res Int. 2023;169:112853.
35. Chen HJ, Dai FJ, Chen CY, Fan SL, Zheng JH, Chau CF, et al. Effects of molecular weight fraction on antioxidation capacity of rice protein hydrolysates. Sci Rep. 2023;13(1):3464.

36. Cione E, Fazio A, Curcio R, Tucci P, Lauria G, Cappello AR, et al. Resistant starches and non-communicable disease: a focus on mediterranean diet. Foods. 2021;10(9):2062.
37. Chen Y, She Y, Zhang R, Wang J, Zhang X, Gou X. Use of starch-based fat replacers in foods as a strategy to reduce dietary intake of fat and risk of metabolic diseases. Food Sci Nutr. 2020;8(1):16–22.
38. Huang M, Mehany T, Xie W, Liu X, Guo S, Peng X. Use of food carbohydrates towards the innovation of plant-based meat analogs. Trends Food Sci Technol. 2022;129:155–63.
39. Xie Y, Cai L, Zhao D, Liu H, Xu X, Zhou G, et al. Real meat and plant-based meat analogues have different in vitro protein digestibility properties. Food Chem. 2022;387:132917.
40. Yang Y, Zheng Y, Ma W, Zhang Y, Sun C, Fang Y. Meat and plant-based meat analogs: nutritional profile and in vitro digestion comparison. Food Hydrocoll. 2023;143:108886.
41. Lyu Z, Sala G, Scholten E. Melting properties of vegan cheese: effect of emulsion and protein addition on the thermal behaviour of starch gels. Food Hydrocoll. 2023;144:108917.
42. Vu G, Zhou H, McClements DJ. Impact of cooking method on properties of beef and plant-based burgers: appearance, texture, thermal properties, and shrinkage. J Agric Food Res. 2022;9:100355.
43. Eshag Osman MF, Mohamed AA, Mohamed Ahmed IA, Alamri MS, Al Juhaimi FY, Hussain S, et al. Acetylated corn starch as a fat replacer: effect on physiochemical, textural, and sensory attributes of beef patties during frozen storage. Food Chem. 2022;388:132988.
44. Dharaiya CN, Jana AH, Aparnathi KD. Functionality of Mozzarella cheese analogues prepared using varying protein sources as influenced by refrigerated storage. J Food Sci Technol. 2019;56(12):5243–52.
45. Dinali M, Liyanage R, Silva M, Newman L, Ahikari B, Wijesekara I, et al. Fibrous structure in plant-based meat: high-moisture extrusion factors and sensory attributes in production and storage. Food Rev Intl. 2024;40:2940.
46. Jimenez-Colmenero F, Cofrades S, Herrero AM, Solas MT, Ruiz-Capillas C. Konjac gel for use as potential fat analogue for healthier meat product development: effect of chilled and frozen storage. Food Hydrocoll. 2013;30(1):351–7.
47. Ahsan M, Ali TM, Hasnain A. Use of oxidized potato starch as simultaneous fat and casein replacer in analogue mozzarella cheese-II: impact on functional and sensory properties of cheese. Food Hydrocoll. 2023;142:108810.
48. Zehra N, Ali TM. Potential of hydroxypropylated corn and sorghum starch extrudates as simultaneous fat and casein mimetics in analogue mozzarella cheese: insights from functional, rheological, textural, and sensory analysis. Starch - Stärke. 2024;76(7–8):2300209.
49. Godschalk-Broers L, Sala G, Scholten E. Meat analogues: relating structure to texture and sensory perception. Food Secur. 2022;11(15):2227.
50. Hamaker BR. Current and future challenges in starch research. Curr Opin Food Sci. 2021;40:46–50.
51. Xia Y, Qian J, Zhao Y, Zheng B, Wei K, Peng B, et al. Effects of food components and processing parameters on plant-based meat texture formation and evaluation methods. J Texture Stud. 2023;54(3):394–409.
52. McClements DJ, Grossmann L. The science of plant-based foods: constructing next-generation meat, fish, milk, and egg analogs. Compr Rev Food Sci Food Saf. 2021;20(4):4049–100.
53. McClements DJ. Development of next-generation nutritionally fortified plant-based milk substitutes: structural design principles. Food Secur. 2020;9(4):421.
54. Sethi S, Tyagi SK, Anurag RK. Plant-based milk alternatives an emerging segment of functional beverages: a review. J Food Sci Technol. 2016;53(9):3408–23.
55. Tangyu M, Muller J, Bolten CJ, Wittmann C. Fermentation of plant-based milk alternatives for improved flavour and nutritional value. Appl Microbiol Biotechnol. 2019;103(23–24):9263–75.
56. Compart J, Singh A, Fettke J, Apriyanto A. Customizing starch properties: a review of starch modifications and their applications. Polymers (Basel). 2023;15(16):3491.

57. Musa-veloso K, Juana J. Regulation and labeling of plant-based beverages and simulated meat, poultry, and egg products in Canada and the United States. Cereal Foods World. 2020;65(4):44–6.
58. Han Z, Shi R, Sun DW. Effects of novel physical processing techniques on the multi-structures of starch. Trends Food Sci Technol. 2020;97(October 2019):126–35.
59. Maniglia BC, Castanha N, Rojas ML, Augusto PE. Emerging technologies to enhance starch performance. Curr Opin Food Sci. 2021;37:26–36.
60. Rodriguez-Ruiz JD, Rodriguez-Sandoval E, Hernández MS. Influence of modified cassava starches on the rheological and structural properties of a fermented soy beverage during storage. Starch/Staerke. 2024;2400003:1–9.
61. Grossmann L, McClements DJ. The science of plant-based foods: approaches to create nutritious and sustainable plant-based cheese analogs. Trends Food Sci Technol. 2021;118(PA):207–29.
62. Grasso N, Roos YH, Crowley SV, Arendt EK, O'Mahony JA. Composition and physicochemical properties of commercial plant-based block-style products as alternatives to cheese. Future Foods. 2021;4:100048.
63. Sridhar K, Bouhallab S, Croguennec T, Renard D, Lechevalier V. Recent trends in design of healthier plant-based alternatives: nutritional profile, gastrointestinal digestion, and consumer perception. Crit Rev Food Sci Nutr. 2023;63(30):10483–98.

Chapter 10
Clean Label Starches

Nicolas M. Bogdanoff, Omayra B. Ferreiro, and Fátima Díaz

10.1 Introduction

Foods and beverages containing clean, simple ingredients that consumers can recognize are not a new phenomenon, yet they have gained significant attention in recent years as health-conscious consumers increasingly want to know exactly what they are eating. "Clean label foods" implies that products are (1) free of synthetic additives or preservatives, (2) made from raw food materials that are easily identifiable by consumers, and (3) produced using traditional food processing methods [1].

The movement to apply clean labels to food products—by meeting essential requirements such as using natural and minimally processed materials—originated in the United Kingdom during the 1990s [2]. Today, consumers no longer simply choose "healthy foods," they examine the ingredients in their food and consider the product's environmental impact [3, 4]. Consequently, the clean label concept now extends throughout the entire food processing industry.

Clean label is a consumer-driven movement demanding a return to real food through honesty and transparency. Foods containing natural, familiar, and simple

N. M. Bogdanoff (✉)
Universidad Paraguayo Alemana. Facultad de Ciencias de la Ingeniería, San Lorenzo, Paraguay
e-mail: nicolas.bogdanoff@upa.edu.py

O. B. Ferreiro
Universidad Nacional de Asunción, Facultad de Ciencias Químicas, Departamento de Ingeniería y Tecnología de Alimentos, San Lorenzo, Paraguay

Universidad Nacional de Asunción, Facultad Politécnica, Grupo de Investigación en Bio y Materiales, San Lorenzo, Paraguay

F. Díaz
Universidad Nacional de Asunción, Facultad de Ciencias Químicas, Departamento de Ingeniería y Tecnología de Alimentos, San Lorenzo, Paraguay

G. Ayala Valencia (ed.), *Starch: Progress in Food Applications*,
https://doi.org/10.1007/978-3-031-98732-8_10

ingredients that are easy to recognize, understand, and pronounce should be free of synthetic chemicals [5]. Thus, aspects such as "free of additives," "minimally processed," and a "simple ingredient list" are considered essential requirements for clean label foods [3, 6, 7]. In fact, many meticulous consumers place even greater importance on clean label foods than on those from well-known brands [8]. According to a Nielsen survey, sales of clean label products continue to grow, while those of traditional products are on the decline.

Taking all these definitions into account, it can be said that the interpretation of "clean label" is not objective. It depends on the consumer's familiarity with food ingredients and/or production methods. Additionally, there is no specific section in supermarkets dedicated to clean label foods, nor do products carry the words "clean label" on their packaging. However, most definitions agree that a clean label should feature a short, simple ingredient list with terms such as "natural," "organic," or "free from additives." Conversely, terms that evoke chemicals or E numbers should be avoided. Clean label is a global trend that is expanding into new segments and geographies.

The COVID-19 pandemic has increased the number of consumers who are conscious of their health and overall wellbeing. Consumers are more connected, informed, and concerned about their purchases. It is essential to highlight that consumer expectations for clean label products revolve around three core elements:

(a) A list of natural, familiar, and simple ingredients
(b) Nutrient-dense ingredients
(c) Sustainable production practices that consider animal welfare without compromising the needs of future generations

Table 10.1 summarizes the terms that should be included and those that should be avoided in clean label foods.

Below are definitions for some of the terms in the Table 10.1:

Table 10.1 Terms that should be included and those that should be avoided in clean label foods

To include	To avoid
Environmentally sustainable	Artificial
Raw material	Allergenic
All natural/natural	Additives/E-numbers
Organic	Industrially
No antibiotics	Unfamiliar
No-GMO	Chemical-sounding
Free/free from/no	Synthetic
Simple	Preservative
Traditional	
High-fructose syrup	
Shorter ingredient list	
Accepted by consumers	
Locally grown	

- Additive: Substances intentionally added to food to change, modify, or improve its organoleptic characteristics, or to enhance its production process, preservation, or shelf life.
- Non GMO: Free of substances derived from genetically modified organisms.
- Organic: Based on the principles of organic production (as legally regulated), in which chemical fertilizers, synthetic pesticides, or genetically modified organisms are not allowed, thereby respecting the environment and biodiversity.
- Natural: Refers to ingredients that originate in nature, are made truthfully without artifice, mixing, or any additional composition. "Natural food" is understood to be food offered without any alteration to its original composition as it occurs in nature.
- Simple: Based on a recognizable ingredient list, typically containing fewer than 10 ingredients.
- Free from: (Negative) claims indicating that certain ingredients, nutrients, or substances are not present in a food—for example: free from/without/no "artificial colors," "artificial flavors," "fat," "gluten," "lactose," "BPA," "palm oil," "artificial preservatives," "artificial sweeteners," "hormones," "antibiotics," "additives," "monosodium glutamate," "GMOs," and "allergenic compounds."
- Artificial: Synthetic; not natural.

While many factors play a role in the demand for clean label foods, existing research suggests that concerns about health and sustainability—among other issues—motivate consumers to seek these products [9, 10]. The trend away from highly processed ingredients in the food industry is evident in how companies across the retail sector are striving to "clean up" their labels. In the United States, major companies began revising their ingredient lists around 2010 [11, 12]. For example, in 2015 Hershey reformulated its products by replacing sugar derived from sugar beet molasses—a crop predominantly grown from genetically modified seeds—with cane sugar [13]. A year later, Campbell launched its clean label soup line, "Well Yes!", which contained no artificial flavors, colors, or modified starches [14]. Today, clean labels are ubiquitous across multiple food categories, including baked goods, beverages, snacks, ready-to-eat soups, and dairy products [15, 16].

Ingredient "blacklists," compiled by wholesale distributors such as Kroger and Whole Foods, are among the criteria sources for companies developing clean labels [17]. A common approach to clean label reformulation involves either completely removing undesirable ingredients or replacing them with more natural alternatives. The latter process is generally expensive and time-consuming, as these alternatives tend to be more costly, and the resulting products often exhibit less desirable sensory characteristics compared to their original counterparts [18, 19]. In particular, ingredients considered undesirable by consumers are typically those that have been designed or modified to maximize functionality within the food. Consequently, replacing such ingredients (for example, modified corn starch) with natural alternatives (e.g., native corn starch) may lead to increased ingredient usage, higher production costs, and/or potentially inferior sensory attributes [20–23]. The complete removal of these ingredients presents similar challenges [24].

Although consumers express a clear preference for cleaner labels, their behaviors and actions sometimes contradict these preferences—especially when other factors are involved [25]. Two of the most important factors in this context are the sensory characteristics and the price of the product. Sensory attributes such as taste, texture, and appearance are commonly identified as key product characteristics for consumers [26–28]. For instance, consumers are not willing to compromise on taste for health benefits in functional foods [29, 30]. Price is another crucial factor influencing purchasing behavior. Streletskaya et al. [31] demonstrated that an increase in price via taxation can reduce the purchase of unhealthy foods, thereby lowering the intake of certain undesirable nutrients (e.g., calories, cholesterol, etc.). However, in some cases—particularly among higher-income consumers—price may have a smaller impact on food demand compared to non-economic factors. Consequently, although the production costs for clean label foods are higher than those of conventional formulations (and these increased costs are passed on to the consumer), it remains unclear whether premium consumers are willing to pay for clean label products. Moreover, while consumers on average show a greater willingness to pay for clean label products, the higher price may not be sufficient to cover the reformulation costs—similar to the premium-cost dynamics observed with organic foods [9, 32]. The trade-off between label cleanliness, sensory characteristics, and price is of particular interest to companies considering reformulation, yet it remains unclear how these factors interact.

Starch, the primary dietary source of carbohydrates, is a natural and abundant ingredient obtained from roots, seeds, tubers, leaves, and stems. Depending on its source, starch granules can vary in characteristics such as shape, size, structure, and chemical composition [33].

The use of native starches in industrial applications is limited due to their low thermal and shear stability, as well as their tendency to retrograde during refrigeration or freezing—thereby reducing food quality [34–36]. During manufacturing, foods containing starch are subjected to considerable shear forces (e.g., high-speed mixing, pumping, homogenization), which significantly affect the final product quality. Consequently, many studies have concluded that modified starches must be developed to replace native starch in order to overcome these limitations [37]. Generally, small amounts of chemicals or hydrocolloids are added to mitigate the drawbacks of native starches. However, these chemicals are considered "additives" rather than "ingredients" and are not suitable for clean label foods [38].

10.2 Clean Label Modified Starches

Clean label starch (CLS) is defined as a starch that provides the functionality of a modified starch while meeting consumer demands across a wide range of applications due to its versatility and various benefits. In contrast, starches modified physically or enzymatically—as well as native starches—are classified as ingredients rather than additives. Therefore, they can be labeled simply as "starches" from various sources without the use of E numbers [39].

The term "native" applies to starches that are extracted and unprocessed, as opposed to "natural." In the context of starch, "natural" might imply that the entire plant from which the starch is extracted is used in the food [40]. From an industrial perspective, native carbohydrates used as fat replacers may offer some advantages for clean label applications; however, their functional performance can be challenging to adjust given the high specificity of some food systems [41]. Starch possesses a property known as pre-gelatinization, which is highly desirable and does not require chemical compounds during production. After pre-gelatinization, drying the starch prevents the crystallization of amylose and amylopectin, resulting in a glassy structure with rapid hygroscopicity. This starch is capable of absorbing both cold and hot water without the need for heating to develop viscosity [42].

Some starches are used as fat replacers to impart desirable properties—such as mouthfeel, creaminess, appearance, and viscosity—to foods. However, they may also mask the flavor of the food into which they are incorporated. Additionally, they are unable to carry liposoluble flavor components, necessitating the inclusion of emulsifiers [43]. Although their rheological stability may be insufficient for some applications, they perform well as fat replacers in high-moisture foods such as salad dressings, sauces, baked goods, frozen foods, and sausages [44].

However, physically modified starches account for only about 5% of modified food starches worldwide, indicating that further development in this area is needed.

Figure 10.1 illustrates the increase in publications related to clean label starch indexed in the Scopus database until 2024 (Query: (TITLE-ABS-KEY("clean label" AND "starch")). Over the past 10 years, the annual output has increased more than seven-fold. Among the countries producing the most documents on clean label starch foods are China and the United States, while within Europe, Spain and Italy

Fig. 10.1 Published papers on clean label starch at the Scopus database [45]

are the leading contributors. Moreover, the clean label movement has become a strategic focus for food producers, involving food agencies—many of which collaborate with foundations that finance scientific research. This has led the Food Institutional Research Measurement to become the agency with the highest volume of scientific research on clean label to date 45.

10.3 Starch Applications in Clean Label Products

The concept of clean label is recent and has been growing due to consumer interest. In general, it is associated with the concept of introducing natural or minimally processed ingredients from natural raw materials and definitely not involving artificial or synthetic compounds [46]. Nowadays it also includes the impact that the product generates on the environment.

Native starches have limited food industrial application due to their thermal and shear instability and the tendency to retrograde during cooling or freezing, very common processes in the food industry. This clearly affects the food product quality and chemical starch modification has widely proposed to overcome its limitations.

Starchy foods are important for nutrition and energy and to be classified as clean labels, starch must be free from chemical modification, but it can be physical or enzymatically modified which will have an influence on morphological, thermal, rheological, and pasting properties [47, 48].

The presence of E-numbers codes or unfamiliar ingredients listed on the label has begun to negatively affect the consumers' perceptions and preferences about the food. This is generally related to the consumer worry about the negative effect that these unfamiliar ingredients could have on health and environment. Some clean label starch applications are described below related to starch non-chemical modification for use as emulsion stabilization, low-glycemic foods and in plant-based foods.

10.4 Clean Label Starch and Its Role in Emulsion Stabilization

Maruyama et al. [11] performed an essay to investigate the factors that affect acceptability and the consumers' perceived naturalness of food ingredients in yogurt. Sugars, stabilizers/thickening agents (among them corn starch), preservatives and coloring agents were selected. The results suggest that stabilizers/thickening agents have effects for clean label reformulations and for ingredient acceptability is more important its source than its functionality. In this way, the authors considered that the functionality of stabilizers/thickening agents may not be as clear as that of coloring or flavors in food lending to the decrease the consumer acceptance of these ingredients.

Gluten-free bread is a challenge for the food industry, since gluten is the ingredient responsible for the characteristics viscoelastic properties of conventional wheat doughs. Based on this, non-gluten cereal-based flours, legumes as well as a large variety of other grains and seeds that could be used as sources of protein, fiber, and minerals. However, mimic the viscoelastic properties of gluten towards the development of high-quality products ingredients such as cassava, corn and potato starch, eggs, sunflower seeds, psyllium, hydroxypropylmethylcellulose and xanthan gum have been proposed. These ingredients with high water-binding ability, film-forming and structure-forming capacities, emulsifiers combining new processing technologies have allowed to improve the quality of gluten-free products [46].

Chemically modified starch has been widely proposed for preparing Pickering emulsions [49–52]. The use of octenyl succinic anhydride (OSA) is the most common modification performed for starch, which consists of esterification, resulting in a hydrophobized starch [51, 52]. The emulsions prepared by the Pickering mechanism are stabilized by solid particles (such as protein aggregates and carbohydrates), which accumulate at the interface between two immiscible liquids and avoid the coalescence of droplets [49, 53]. Although it is considered that native or unmodified starch has less functional and physicochemical characteristics to be used in different applications such as emulsifiers, many authors have demonstrated the application of native starch from different sources in Pickering emulsions without modification or using physical modifications [53–56].

Physical modification of starch is an alternative to conventional stabilizers used, most of them synthetic ones [49, 55]. Rice starch showed better performance than others, such as waxy maize, wheat, and potato starch, which was related to the starch size [54]. It has been confirmed that smaller particles have more efficient emulsifying abilities than larger ones, influencing also on the emulsion droplet sizes [57]. In this way, Guida et al. [53] used nanoprecipitation assisted by an ultrasonic process to obtain cassava starch nanoparticles to produce coalescence-stable emulsions. This approach contributes to the use of natural ingredients and clean label products. Besides the starch size, other factors such as starch concentration have been shown to influence the increase of packing density of granules at the interface and the diminishing of droplet size [54, 57]. Saari et al. [57] concluded that for systems consisting of starch mixture, the particle size alone is insufficient to decide which starch type would dominate at the interface of the emulsion droplets and that the droplet size was more affected for the emulsion time for larger starch granules. Saari et al. [57] also observed that emulsifying capacity was less dependent on the starch type and more on the starch concentration.

Ge et al. [58] proposed an extrusion process for modifying the physicochemical properties and starch structure of highland barley flour. Extrusion resulted in an increase in nutritional properties (such as soluble dietary and β-glucan).

10.5 Clean Label Starch and Its Role in Low-glycemic Foods

The glucose produced from starch metabolism serves primarily as a substrate for the brain and red blood cells because of these starchy foods are considered to provide between 50% and 70% of the energy in a typical human diet [59]. The digestibility of starch is a key determinant of its Glycemic Index (GI), which reflects the rate at which glucose is released into the bloodstream. In this way, the digestibility of starch can be classified as rapidly digested starch (RDS), slowly digested starch (SDS) and resistant starch (RS) depending on the time that it takes to be digested. RDS is digested after 20 min, SDS between 20 and 120 min and RS more than 120 min [60]. Rapid digestion results in a high GI, which in turn increases the risk of overweight, obesity, and related conditions such as type II diabetes, while the incorporation of RS as a functional ingredient with prebiotic effects is highly advantageous [59, 60].

For native starch, it has been reported that the freeze-thawing cycles affect the characteristics of starch gel. During freezing, the ice formation increases the starch concentration accelerating the starch retrogradation, meanwhile during thawing, the melted ice is separated from the gel network by the syneresis phenomenon. This has been related to the starch granule surface formed during mild heating that avoids the release of water during the freeze-thawing cycles conferring freeze-thaw stability [61].

The main difference between the starch sources is related to the amylose and amylopectin content, its size and the presence of surface protein and lipid and this influences the performance of starch in different applications [54, 56]. A high content of amylose in corn starch has been reported to induce rapid retrogradation upon storage, accelerating the release of water during the freeze-thawing process [61]. Blending different types of starch can help to overcome the limitations of each native starch used separately [48].

Physical modification of starch by heat, moisture, radiation or shear have shown to cause a molecular rearrangement transforming RDS to SDS [62]. Heat treatment could be a strategy to modify the starch into functional ingredients without any chemical modification. Heat-moisture treatment (HMT) of potato starch has proved to cause physical crosslinking which allowed the main structure to remain intact with no significant swelling [59]. HMT also has demonstrated to increase the formation of resistant starch [63]. The way to perform the HMT also affects the characteristics of starch. Repeated heat-moisture treatment (RHMT) of starch showed greater advantages in practical applications compared to a continuous heat-moisture treatment (CHMT), which could be attributed to the cooling process in the former process that enhanced the linkage between the recombinant amylose/amylopectin [64]. Starch treated by RHMT also showed higher relative crystalline degree and gelatinization transition temperatures [64]. Huang et al. [65] found that the content of SDS and thermo-stable SDS of sweet potato starch is more influenced by the cycling times than the heat-moisture treatment time. In this way, a significant increase of SDS and thermo-stable SDS content was reached after a RHMT performed 3 times. Yassaroh et al. [59] tested a combination of heat-moisture treatment of potato starch and linoleic acid for improving the stability of the starch granules towards heating and shearing, which was related to less swelling.

Da Cunha et al. [60] evaluated the addition of lyophilized purple-fleshed sweet potato powder on the characteristics of Greek yogurt. The sweet potato was heat-treated in water at 80 °C, processed into puree and lyophilized. The powder presented a total content of starch (TS) of 53.27%, which RDS, SDS and RS were 75.44%, 7.74% and 16.82%, respectively. Meanwhile, the non-treated sample contained 53.14% of TS and 75.35%, 21.54%, and 2.77% of RDS, SDS and RS, respectively. The authors concluded that the heat treatment improved the content of RS due to the denaturalization of amylases, which reduces starch hydrolysis and promotes a debranching of the α-(1, 6) bond of amylopectin that is converted into small-chain linear polysaccharides. The incorporation of lyophilized sweet potato powder into Greek yogurt also contributes to obtaining a firm yogurt with greater chewability, reduced syneresis and rich in bioactive components.

Recently, Park et al. [61] proposed a physical modification of starch by mild heating of a hydroalcoholic suspension, achieving an increase in gelatinization and pasting temperatures and an improvement in stability against shear and heat. The authors also observed that the improvement in gel properties was more pronounced in potato starch than in tapioca and corn starches. Potato starch showed a better performance when heat-treated for stabilizing during freeze-thawing cycles; meanwhile, as native potato starch, it had no emulsifying ability in Pickering emulsions, even in high concentrations [54].

10.6 Clean Label Starch and Its Role in Plant-based Foods

The growing demand for more natural products is due to the fact that there are consumers who demand more natural and organic foods, and minimally processed, clean label products, without allergens and/or without additives, are the most requested. Consumers are primarily looking to consume more plant-based food products.

Native vegetable starches and grains can be used as alternatives to replace texturing and thickening agents and emulsifiers mono- and di-glycerides, fulfilling functions to improve texture, emulsification and as stabilizer [66]. These native starches are obtained by separating the natural starch from cereals or tubers (such as cassava, rice, corn, and potatoes) and can be used directly in the production of certain foods such as noodles [40].

Native quinoa starch was reported to have better emulsifying capacity than other small-sized starches, such as rice starch and amaranth starch. It has a better emulsifying capacity than other small-sized starches, such as rice starch and amaranth starch. A dry heat treatment was even used to modify quinoa starch; However, the modified starch did not have good emulsifying properties [52, 67].

For plant-based mayonnaise, modified starch has been proposed as an ingredient for improving the texture thickener [68]. Dan et al. [69] performed a physical modification of wheat starch by an annealing process that increased pasting viscosity and decreased swelling power and water solubility index of annealed wheat starch compared to native wheat starch. A low-fat mayonnaise with enhanced viscoelastic properties and had brighter color was prepared with annealed starch and quercetin-loaded annealed starch by replacing fat content at 50% demonstrated to have higher emulsifying stability than a full-fat mayonnaise providing a healthier alternative to conventional mayonnaise [69].

10.7 Challenges and Limitations

Despite the promising advances and growing consumer demand for clean label starches, several challenges and limitations remain that must be addressed to fully realize their potential in food applications. First, while physical and enzymatic modifications offer a viable alternative to chemical processes, they often fall short in replicating the full functionality and stability provided by their chemically modified counterparts. This gap poses significant technical hurdles, particularly in applications requiring precise rheological and sensory properties.

Economic factors also play a critical role. The reformulation process necessary to meet clean label standards frequently incurs higher production costs, which may not be fully recoverable through premium pricing. Additionally, consumer behavior sometimes diverges from stated preferences; while there is a growing willingness to embrace clean label products, the premium price associated with these formulations can limit broader market acceptance.

Furthermore, the variability inherent in natural starch sources—stemming from differences in botanical origin, amylose-to-amylopectin ratio, and granule morphology—introduces additional complexity in standardizing performance across diverse food matrices. These inconsistencies necessitate further research and development to optimize processing conditions and to tailor starch functionality for specific applications without compromising product quality.

10.8 Conclusions

As the clean label movement expands globally, the lack of uniform regulatory definitions and labeling standards creates further ambiguity for both manufacturers and consumers. Establishing clear guidelines and fostering collaborative efforts between researchers, industry, and regulatory bodies will be essential to overcome these challenges.

In conclusion, while clean label starches represent a significant stride toward more transparent and sustainable food production, addressing these technical, economic, and regulatory challenges will be crucial for their successful integration into mainstream food systems. Future research should focus on enhancing the functional performance of physically and enzymatically modified starches, reducing production costs, and developing standardized criteria to support the clean label ethos without compromising product quality or consumer trust.

Acknowledgements The authors acknowledge the support given by the Universidad Paraguayo Alemana and the Universidad Nacional de Asunci6n for the development of this study.

Conflicts of Interest The authors declare no conflict of interest.

References

1. Ryu YA, Lee JS. Clean label guideline for entry into UK and EU agro-food markets. Food Ind Nutr. 2018;23(2):20–6.
2. Baines D, Seal R. Natural food additives, ingredients and flavourings. Elsevier; 2012. p. 1–22.
3. Aschemann-Witzel J, Varela P, Peschel AO. Consumers' categorization of food ingredients: do consumers perceive them as 'clean label' producers expect? An exploration with projective mapping. Food Qual Prefer. 2019;71:117–28.
4. Seo S, Ahn HK, Jeong J, Moon J. Consumers' attitude toward sustainable food products: ingredients vs. packaging. Sustainability. 2016;8(10):1073.
5. Hutt CA, Sloan A. Coming clean: what clean label means for consumers and industry. In: Global food forum clean label conference. Itasca: Westin Hotel; 2015. p. 5–8.
6. Busken DF. Cleaning it up—what is a clean label ingredient? Cereal Foods World. 2015;60(2):112–3.
7. Ingredion. The clean label guide in Europe [Internet]. 2014. Disponible en: http://www.alimentatec.com/wp-content/uploads/2014/10/The-Clean-Label-Guide-To-Europe.pdf.
8. De Oliveira Do Nascimento K, Do Nascimento Dias Paes S, Maria Augusta I. A review «Clean Labeling»: applications of natural ingredients in bakery products. J Food Nutr Res. 2018;6(5):285–94.
9. Asioli D, Aschemann-Witzel J, Caputo V, Vecchio R, Annunziata A, Naes T. Making sense of the "clean label" trends: a review of consumer food choice behavior and discussion of industry implications. Food Res Int. 2017;99:58–71.
10. Honkanen P, Verplanken B, Olsen S. Ethical values and motives driving organic food choice. J Consum Behav. 2006;5:420–30.
11. Maruyama S, Streletskaya NA, Lim J. Clean label: why this ingredient but not that one? Food Qual Prefer. 2021;87:104062.
12. Berenstein N. Clean dirty little secret. The Counter [Internet]. 8 de julio de 2018; Disponible en: https://thecounter.org/clean-label-dirty-little-secret/.
13. Charles D. As big candy ditches GMOs, sugar beet farmers hit a sour patch. Food Thought [Internet]. 8 de julio de 2016; Disponible en: https://www.npr.org/sections/thesalt/2016/05/12/477793556/as-big-candy-ditches-gmos-sugar-beetfarmers-hit-sour-patch.
14. Schouten R. Campbell Soup launches new clean label soup line. Food Bus News [Internet]. 8 de julio de 2016; Disponible en: https://www.foodbusinessnews.net/articles/7353-campbell-soup-launches-new-clean-label-soup-line.
15. Shelke K. Clearing up clean label confusion. Food Technol. 2020;74:40–51.
16. Onward DJ. Upward CLTSNS of S. 2019. Disponible en: https://www.fooddive.com/news/onward-and-upward-clean-label-trend-shows-no-signs-of-slowing/545738/.

17. Out WEGCSR, Mean? BWDI. 2017. Disponible en: https://www.foodnavigator-usa.com/Article/2017/06/13/GoCleanLabel-certified-scheme-rolls-out-but-what-does-it-mean.
18. Van Gunst A, Roodenburg AJ, Steenhuis IH. Reformulation as an integrated approach of four disciplines: a qualitative study with food companies. Food Secur. 2018;7:64.
19. Frans H. Labelling of food: a challenge for many. Vet Glas. 2017;71:16–23.
20. Krawiec S. Can clean-label starches withstand freezing and high-heat applications? Nutr Outlook. 2019;22:62.
21. Michail N. Modified maize starch brings cost savings of 15%, says Ingredion [Internet]. 2017. Disponible en: https://www.foodnavigator.com/Article/2017/05/19/Modified-maize-starch-brings-cost-savings-of-15-says-Ingredion.
22. Miyazaki M, Hung P, Maeda T, Morita N. Recent advances in application of modified starches for breadmaking. Trends Food Sci Technol. 2006;17:591–9.
23. Association CR. Corn starch. 11th ed. West Lafayette: Corn Refiners Association; 2006.
24. Poinski M. Cleanliness conundrums: the challenges in clean label reformulations [Internet]. 2017. Disponible en: https://www.fooddive.com/news/cleanliness-conundrums-the-challenges-in-clean-label-reformulations/447098/.
25. Davies MAP, Wright LT. The importance of labelling examined in food marketing. Eur J Mark. 1994;28:57–67.
26. Herrera-Corredor JA, Saidu JEP, Khachatryan A, Prinyawiwatkul W, Carballo-Carballo A, Zepeda-Bautista R. Identifying drivers for consumer acceptance and purchase intent of corn tortilla. J Food Sci. 2007;72:727–31.
27. Grunert KG, Bech-Larsen T, Bredahl L. Three issues in consumer quality perception and acceptance of dairy products. Int Dairy J. 2000;10:575–84.
28. Imram N. The role of visual cues in consumer perception and acceptance of a food product. Nutr Food Sci. 1999;99:224–30.
29. Lyly M, Roininen K, Honkapaa K, Poutanen K, Lahteenmaki L. Factors influencing consumers' willingness to use beverages and ready-to-eat frozen soups containing oat beta-glucan in Finland, France and Sweden. Food Qual Prefer. 2007;18:242–55.
30. Verbeke W. Functional foods: consumer willingness to compromise on taste for health? Food Qual Prefer. 2006;17:126–31.
31. Streletskaya NA, Rusevichientong P, Amatyakul W, Kaiser HM. Taxes, subsidies, and advertising efficacy in changing eating behavior: an experimental study. Appl Econ Perspect Policy. 2013;36:146–74.
32. Vecchio R, Loo EJ, Annunziata A. Consumers' willingness to pay for conventional, organic and functional yogurt: evidence from experimental auctions. Int J Consum Stud. 2016;40:368–78.
33. Smith AM. The biosynthesis of starch granules. Biomacromolecules. 2001;2(2):335–41.
34. Arocas A, Sanz T, Fiszman SM. Clean label starches as thickeners in white sauces. Shearing, heating and freeze/thaw stability. Food Hydrocoll. 2009;23(8):2031–7.
35. Karam LB, Grossmann MVE, Silva RSS, Ferrero C, Zaritzky NE. Gel textural characteristics of corn, cassava and yam starch blends: a mixture surface response methodology approach. Starch/Staerke. 2005;57(2):62–70.
36. Kokini J, Wang CF, Huang H, Shrimanker S. Constitutive models of foods. J Texture Stud. 1995;26(4):421–55.
37. Zia-ud-Din X, Xiong H, Fei P. Physical and chemical modification of starches: a review. Crit Rev Food Sci Nutr. 2017;57(12):2691–705.
38. Park S, Kim YR. Development of non-linear prediction model for starch blending. LWT. 2023;190:115567.
39. Radeloff MA. Clean label—Starches and their functional diversity. Sugar Ind. 2016;141(4):209–15.
40. McDonagh P. Native, modified and clean label starches in foods and beverages. In: Natural food additives, ingredients and flavourings [Internet]. Elsevier; 2012 [citado 12 de noviembre de 2024]. p. 162–74. Disponible en: https://linkinghub.elsevier.com/retrieve/pii/B9781845698119500074.

41. Peng X, Yao Y. Carbohydrates as fat replacers. Annu Rev Food Sci Technol. 2017;8:331–51.
42. Rahmati NF, Mazaheri Tehrani M, Daneshvar K, Koocheki A. Influence of selected gums and pregelatinized corn starch on reduced fat mayonnaise: modeling of properties by central composite design. Food Biophys. 2014;10:39–50.
43. Fellows PJ. Properties of food and principles of processing. In: Food processing technology. Cambridge: Woodhead Publishing; 2017.
44. O'Connor TP, O'Brien NM. Cheese: low-fat and reduced-fat cheese. En: fat replacers. Amsterdam: Elsevier; 2016.
45. Scopus database: search results for TITLE-ABS-KEY("clean label" AND "starch"), *Scopus (Elsevier B.V.)*, Accessed 10 Oct 2024, https://www.scopus.com(2024)
46. Papagianni E, Kotsiou K, Matsakidou A, Biliaderis CG, Lazaridou A. Development of "clean label" gluten-free breads fortified with flaxseed slurry and sesame cake: implications on batter rheology, bread quality and shelf life. Food Hydrocoll. 2024;150:109734.
47. Sridhar K, Sharma M, Inbaraj BS. Functional clean-label starch: sustainable production technologies and food applications. Starch - Stärke. 2024;76(11–12):2400157.
48. Park S, Kim YR. Clean label starch: production, physicochemical characteristics, and industrial applications. Food Sci Biotechnol. 2021;30(1):1–17.
49. Yang Y, Fang Z, Chen X, Zhang W, Xie Y, Chen Y, et al. An overview of pickering emulsions: solid-particle materials, classification, morphology, and applications. Front Pharmacol. 2017;8:287.
50. Ye F, Miao M, Jiang B, Campanella OH, Jin Z, Zhang T. Elucidation of stabilizing oil-in-water Pickering emulsion with different modified maize starch-based nanoparticles. Food Chem. 2017;229:152–8.
51. Agama-Acevedo E, Bello-Perez LA. Starch as an emulsions stability: the case of octenyl succinic anhydride (OSA) starch. Curr Opin Food Sci. 2017;13:78–83.
52. Marefati A, Wiege B, Haase NU, Matos M, Rayner M. Pickering emulsifiers based on hydrophobically modified small granular starches—Part I: manufacturing and physico-chemical characterization. Carbohydr Polym. 2017;175:473–83.
53. Guida C, Aguiar AC, Magalhães AER, Soares MG, Cunha RL. Impact of ultrasound process on cassava starch nanoparticles and pickering emulsions stability. Food Res Int. 2024;192:114810.
54. Li C, Li Y, Sun P, Yang C. Pickering emulsions stabilized by native starch granules. Colloids Surf Physicochem Eng Asp. 2013;431:142–9.
55. Zhu F. Starch based Pickering emulsions: fabrication, properties, and applications. Trends Food Sci Technol. 2019;85:129–37.
56. Kasprzak MM, Macnaughtan W, Harding S, Wilde P, Wolf B. Stabilisation of oil-in-water emulsions with non-chemical modified gelatinised starch. Food Hydrocoll. 2018;81:409–18.
57. Saari H, Rayner M, Wahlgren M. Effects of starch granules differing in size and morphology from different botanical sources and their mixtures on the characteristics of pickering emulsions. Food Hydrocoll. 2019;89:844–55.
58. Ge F, Sun Y, Yang C, Cheng W, Wang Z, Xia X, et al. Exploring the relationship between starch structure and physicochemical properties: the impact of extrusion on highland barley flour. Food Res Int. 2024;183:114226.
59. Yassaroh Y, Woortman AJJ, Loos K. A new way to improve physicochemical properties of potato starch. Carbohydr Polym. 2019;204:1–8.
60. Cunha Júnior PCD, Silva MTDC, Barbosa MIMJ, Ferreira EHDR. Application of lyophilized purple-fleshed sweet potato powder as a multifunctional ingredient in Greek yogurt. Ciênc Rural. 2024;54(3):e20220688.
61. Park H, Cho D, Chung H, Lim S. Enhanced gelling property and freeze–thaw stability of potato, tapioca and corn starches modified by mild heating in aqueous ethanol solution. J Sci Food Agric. 2024;104(12):7228–37.
62. Wang H, Zhang B, Chen L, Li X. Understanding the structure and digestibility of heat-moisture treated starch. Int J Biol Macromol. 2016;88:1–8.

63. Brahma B, Sit N. Physicochemical properties and digestibility of heat moisture–treated potato starches for different treatment conditions. Potato Res. 2020;63(3):367–83.
64. Zhang B, Saleh ASM, Su C, Gong B, Zhao K, Zhang G, et al. The molecular structure, morphology, and physicochemical property and digestibility of potato starch after repeated and continuous heat–moisture treatment. J Food Sci. 2020;85(12):4215–24.
65. Huang TT, Zhou DN, Jin ZY, Xu XM, Chen HQ. Effect of repeated heat-moisture treatments on digestibility, physicochemical and structural properties of sweet potato starch. Food Hydrocoll. 2016;54:202–10.
66. Cano-Lamadrid M, Valverde JM, Lipan L, Carbonell-Barrachina ÁA, Sendra E. Introduction. In: Galanakis CM, editor. The age of clean label foods [Internet]. Cham: Springer International Publishing; 2022. p. 1–35. Disponible en: https://doi.org/10.1007/978-3-030-96698-0_1.
67. Rayner M, Sjöö M, Timgren A, Dejmek P. Quinoa starch granules as stabilizing particles for production of Pickering emulsions. Faraday Discuss. 2012;158:139–55; discussion 239–266.
68. Pereira RC, Bourbon AI, Azevedo AG, Figueira D, Castelo-Branco D, Rocha C, et al. Mayonnaise produced by ultrasound-assisted emulsification using plant-based and "clean label" ingredients. Food Biosci. 2024;61:104847.
69. Dan Y, Baek Y, Jeong EW, Lee HG. Development of a novel fat reduction system with quercetin-loaded annealed wheat starch for enhanced emulsifying and oxidative stability in low-fat mayonnaise. J Food Eng. 2024;364:111812.

Index

G. Ayala Valencia (ed.), *Starch: Progress in Food Applications*, https://doi.org/10.1007/978-3-031-98732-8

GPSR Compliance

The European Union's (EU) General Product Safety Regulation (GPSR) is a set of rules that requires consumer products to be safe and our obligations to ensure this.

If you have any concerns about our products, you can contact us on ProductSafety@springernature.com

In case Publisher is established outside the EU, the EU authorized representative is:

Springer Nature Customer Service Center GmbH
Europaplatz 3
69115 Heidelberg, Germany

Batch number: 10755332

Printed by Printforce, the Netherlands